Lecture Notes in Mechanical Engineering

About this Series

Lecture Notes in Mechanical Engineering (LNME) publishes the latest developments in Mechanical Engineering—quickly, informally and with high quality. Original research reported in proceedings and post-proceedings represents the core of LNME. Also considered for publication are monographs, contributed volumes and lecture notes of exceptionally high quality and interest. Volumes published in LNME embrace all aspects, subfields and new challenges of mechanical engineering. Topics in the series include:

- Engineering Design
- Machinery and Machine Elements
- Mechanical Structures and Stress Analysis
- Automotive Engineering
- Engine Technology
- Aerospace Technology and Astronautics
- Nanotechnology and Microengineering
- Control, Robotics, Mechatronics
- MEMS
- Theoretical and Applied Mechanics
- Dynamical Systems, Control
- Fluid Mechanics
- Engineering Thermodynamics, Heat and Mass Transfer
- Manufacturing
- Precision Engineering, Instrumentation, Measurement
- Materials Engineering
- Tribology and Surface Technology

More information about this series at http://www.springer.com/series/11236

Alexander Kravcov · Elena B. Cherepetskaya
Vaclav Pospichal
Editors

Durability of Critical Infrastructure, Monitoring and Testing

Proceedings of the ICDCF 2016

Springer

Editors
Alexander Kravcov
Department of Construction Technology
Czech Technical University
Prague
Czech Republic

Vaclav Pospichal
Faculty of Civil Engineering
Czech Technical University in Prague
Prague
Czech Republic

Elena B. Cherepetskaya
MISiS
National University of Science
and Technology
Moscow
Russia

ISSN 2195-4356 ISSN 2195-4364 (electronic)
Lecture Notes in Mechanical Engineering
ISBN 978-981-10-3246-2 ISBN 978-981-10-3247-9 (eBook)
DOI 10.1007/978-981-10-3247-9

Library of Congress Control Number: 2016959396

Printed on acid-free paper

This Springer imprint is published by Springer Nature
The registered company is Springer Nature Singapore Pte Ltd.
The registered company address is: 152 Beach Road, #22-06/08 Gateway East, Singapore 189721, Singapore

Preface

The International Conference on Durability of Critical Infrastructure, Monitoring and Testing (ICDCF 2016) is a conference intended to be a forum for researchers in all related fields, in addition to those listed under the conference topics and case studies describing practical experiences as well. The conference is aimed at engineers, scientists, field researchers, managers and other specialists involved in the theoretical and practical aspects in the fields of safety, sustainability and durability of the critical infrastructure against impact of all hazards on a project of the objects of critical infrastructure: natural, criminal, terrorist, and accidental, that in fact is a multi-hazard approach for modern security design.

The topics of the conference include risk analysis and risk management; safety material engineering; integrated safety design process; natural disaster management; emergency response; critical infrastructure protection; risk analysis, assessment and management; diagnosing and testing of materials; new materials for safety engineering; modelling and experiments; construction safety and security; safety of users in evacuation; emergency and crisis management; explosive engineering; underground structures and tunnels; new standards on security engineering.

This issue contains selected papers from the ICDCF 2016, which took place in Šatov, a beautiful region of South Moravia, from December 6 to December 12.

Finally, as guest editors, we would like to express our sincere appreciation to the members of the editorial committee and the advisory board for giving an opportunity to work with us. We are grateful to all the authors and reviewers of the manuscripts, and especially Prof. Vladimir Křístek, Conference Chairman, for his valuable advice and his support team from the University of Defence in Brno.

We thank the generous support by National University of Science and Technology MISiS, Willenberg Foundation for their enthusiasm and help with the organization of the conference and we thank the office of the Springer Inc. for assistance with editorial matters.

Elena B. Cherepetskaya
Vaclav Pospichal
Alexander N. Kravcov

Contents

Study of the Internal Structure of Isotropic Pyrolytic Graphite by Broadband Ultrasonic Spectroscopy

Anton S. Bychkov[1(✉)], Elena B. Cherepetskaya[1], Adam Konvalinka[2], Alexander A. Karabutov[1], Alexander Kravcov[2], V.A. Makarov[3], Elena A. Mironova[1,3], and Nikolay A. Morozov[1]

[1] The National University of Science and Technology "MISiS", Leninsky Pr.6, Moscow 119991, Russia
{bychkov, aak}@optoacoustic.ru, echerepetskaya@mail.ru, {lenamironova71, morozov499}@yandex.ru

[2] Faculty of Civil Engineering, Czech Technical University in Prague, Thákurova 7/2077, 166 29 Prague 6 - Dejvice, Czech Republic
{adam.konvalinka, kravtale}@fsv.cz

[3] International Laser Center of M.V. Lomonosov Moscow State University, Moscow, Russia
vamakarov@phys.msu.ru

Abstract. As material for structural components, isotropic pyrolytic graphite (IPG) is widely used in such objects of critical infrastructure as nuclear fission reactors. There is the possibility of using IPG as a facing material for experimental thermonuclear facilities; this is currently being explored. Also, IPG is used as a sealing and anti-friction material in the aerospace industry. It is shown that X-ray tomography is not very suitable for studying the internal structure of IPG produced by chemical vapor deposition because its density is much the same as that of heterogeneities encountered in IPG, such as carbon black inclusions and anisotropic pyrographite crystallites. Broadband ultrasonic spectroscopy has been used to detect pores and carbon deposited layers in IPG samples, 2D images of the internal structure produced.

Keywords: Pyrolytic graphite · X-ray tomography · Wideband ultrasound testing · Internal structure of samples

1 Introduction

Isotropic pyrolytic graphite (IPG) is widely used for manufacturing critical components for the aircraft and aerospace industries; it is a sealing and anti-friction material capable of withstanding high and low temperatures without lubricants at a speed of over 30,000 rpm in hostile environments and in vacuum. IPG is used for manufacturing molds for melting metals; it is a biocompatible material for medical implants, i.e. heart valves, etc. [1]. Graphite is widely used in structural components of nuclear fission

A. Kravcov et al. (eds.), *Durability of Critical Infrastructure, Monitoring and Testing*, Lecture Notes in Mechanical Engineering, DOI 10.1007/978-981-10-3247-9_1

reactors, such as fuel elements and reflectors. It is considered to be suitable for facing experimental thermonuclear facilities [2, 3].

This material is characterized by very high mechanical properties, which are 5 to 10 times higher than those of isostatically pressed graphite at temperatures above 2500°C, which are much higher than those of heat-resistant steels and nickel alloys. Pyrolytic graphite is a non-porous material; a pyrolytic graphite film 0.03 to 0.05 mm thick is impenetrable even to helium at temperatures above 2500°C. The properties of pyrolytic carbon can be improved by adding alloying additives, such as silicon, niobium, boron, cobalt, or hafnium.

The unique physical-mechanical properties of IPG are due to its peculiar structure; the properties can change in a wide range, depending on the parameters of the chemical vapor deposition process.

As a rule, the geometry and physical properties of pyrolytic graphite or other carbon materials change under high temperatures and radiation. Consequently, it is necessary to carry out periodical non-destructive inspections to evaluate the structural and physical-mechanical properties of pyrographite products and their residual life. Moreover, the monitoring of these parameters is important in terms of manufacturing technology selection.

Nowadays, pyrolytic graphite quality is assessed using X-ray diffraction [1] and tomography [1] methods, Raman spectroscopy [3] and scanning/transmission electron microscopy [4–8]. However, these methods are laborious and can be only used to test small-size samples in laboratory. This paper proposes that broadband ultrasonic spectroscopy be used for nondestructive testing of isotropic pyrolytic graphite specimens for such flaws as pores, microcracks, and carbon black inclusions/interlayers.

2 Materials

We investigated samples of isotropic pyrographite produced by chemical vapor deposition. The average density of the isotropic pyrographite was 1.82 g/cm^3. To specify the parameters of the graphite manufacturing process, we studied different

a b c

Fig. 1. Microstructure of isotropic pyrolytic graphite with heterogeneities: a - interlayers; b - separate inclusions; c – cluster of small inclusions (× 160 magnification) temperature regimes of chemical vapor deposition and fluctuating increase in the degree of carbon supersaturation in the gas phase. Under certain conditions, inclusions of anisotropic pyrographite and carbon black (carbon black layers, quite large solitary crystals or clusters of small inclusions) occurred in isotropic pyrolytic graphite. These defects were observed on the surface of some samples (Fig. 1).

3 Methods

The radiographic method proved ineffective for detection of such flaws as cracks, laminations, shells, and cleavages, as well as inclusions of anisotropic pyrographite and carbon black, because the density of isotropic pyrographite is much the same as that of anisotropic pyrographite and carbon black. For example, the density of isotropic pyrographite (ρ_i) varies from 1.75 to 2.10 g/cm^3, that of anisotropic pyrographite (ρ_a) ranges from 1.90 to 2.25 g/cm^3, and the density of carbon black (ρ_{cb}) is 1.76 to 1.95 g/cm^3. Conventional acoustic testing methods are unable to detect such inclusions because of the physical limits.

Consequently, it was proposed that broadband acoustic spectroscopy with a laser ultrasonic source should be used for the diagnostics of the internal structure of the pyrographite. We employed a UDL-02M laser-ultrasonic defectoscope operating in a pulse-echo mode. The defectoscope has a pulse-periodic solid-state laser with wavelength λ = 1064 nm; so it was possible to average over 128 realizations. The pulse energy was 100 μJ; the pulse duration, 10 ns. Laser light was delivered, via a fiber optic cable, to an optoacoustic generator, which excited ultrasonic pulses with pressure amplitude of up to 0.1 MPa and duration of about 70 ns, which has their frequency range from 300 kHz to 20 MHz. The wavelength varied between 0,15 mm and 10 mm within this frequency range. The spatial extent of the pulses was less than 0.2 mm. An

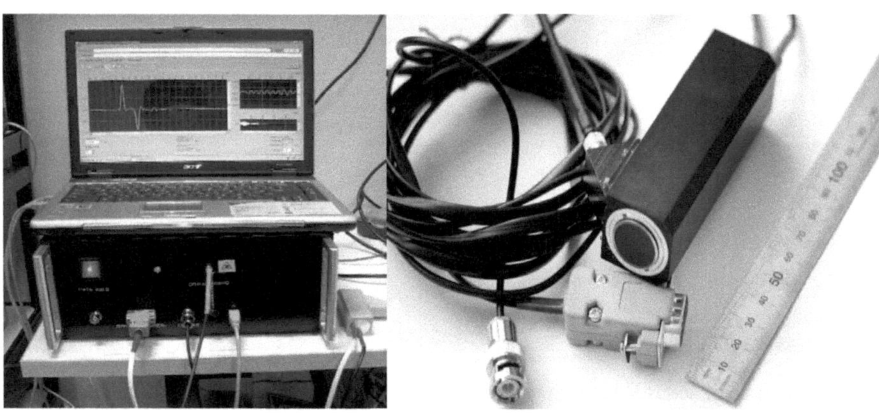

a b

Fig. 2. UDL-2 M defectoscope (a) and opoacoustic transducer (b) external view of UDL-02M defectoscope and optoacoustic generator is shown in Fig. 2.

Ultrasonic pulses (i.e. the reference one and the signals scattered on defects and reflected from front and opposite sides of the sample) were recorded by a damped broadband piezoelectric transducer and converted into electric signals, which were directed, via an amplifier, to a computer. Computer-based signal processing procedure involves deconvolution and filtering of digitized signals.

Given that the sample thickness (h) is known, longitudinal ultrasonic wave velocity c_l is determined from time difference Δt_h between the arrival times of the reference ultrasonic pulse and the pulse reflected from the opposite side of the sample: $c_l = 2h/\Delta t_h$. The distance (h_d) from a defect to the front side of the sample is $h_d = c_l \Delta t_d / 2$, where Δt_d is the delay of the signal reflected from the defect and reflected from front surface. The inaccuracy of calculations, which is determined by the inaccuracy of measurements of h and corresponding time intervals, is less than 0.5 %. The short duration of the signal excited by a generating medium allows defects to be detected at a depth of over 0.2 mm. The "dead" zone is minimal at that.

During reflection of the ultrasonic longitudinal wave pulse from the rear surface of the sample, a shear wave pulse is generated; its arrival is marked by the vertical red line in Fig. 3. The velocity of the shear wave is almost 2 times less than that of the longitudinal wave, so the so the shear wave is observed between the first and second longitudinal reflections in the acoustic recording (1 and 2 in Fig. 3). All modules of elasticity of isotropic solid body can be determined from the measured velocities of these two waves.

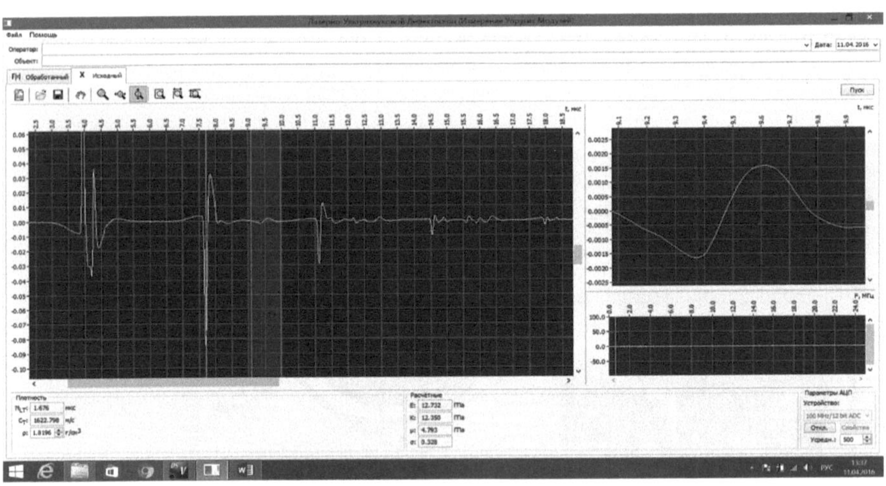

Fig. 3. Example of acoustic recording: 1 - reference signal, 2 - pulse reflected from the back side of the sample, 3 – segment of the acoustic recording with signals scattered by heterogeneities, 4 - signal having travelled through the sample twice, 5 - shear wave pulse generated due to the boundary conditions

4 Results and Discussion

According to the technical specifications, the range of the admissible values of the mechanical parameters of isotropic pyrographite is strictly defined (Table 1).

Table 1. .

Density [ρ], g/cm^3	1.75–2.1
Poisson ratio [ν]	0.32–0.34
Young modulus (dynamic modulus of elongation) [E], HPa	11.9–14.7

Therefore, longitudinal and shear wave velocities c_l and c_t, which were calculated using the formulas

$$c_l = \left(\frac{E(1-\nu)}{\rho(1+\nu)(1-2\nu)}\right)^{0,5}, c_t = \left(\frac{E}{2\rho(1+\nu)}\right)^{0,5} = \left(\frac{G}{\rho}\right)^{0,5} \quad (1)$$

were as follows: c_l = 2847 to 3595 m/s and c_t = 1454 to 1784 m/s. In expressions (1), E is Young's modulus, ν is Poisson's ratio, and G is the shear modulus. Thus, if the velocity values determined from the results of scanning over the surface of a sample fell outside these intervals, it was clear evidence of the presence of defects, therefore these areas of the samples were considered defective. However, if a sample had both inclusions of anisotropic pyrographite crystals and pores/microcracks, the average velocities fell in the above intervals. Therefore, it was necessary to carry out 2D visualization based on the acoustic recordings so as to accurately assess the presence of internal flaws and discontinuities in the test object. The following defects were

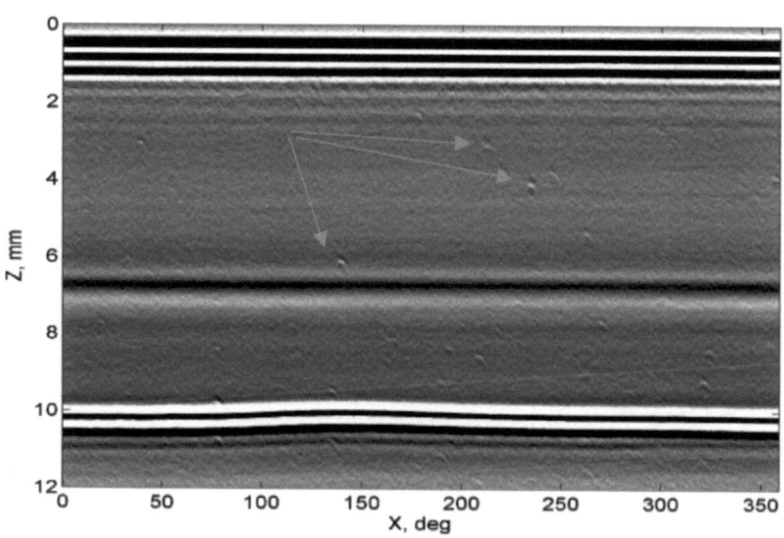

Fig. 4. B-scan of isotropic pyrographite sample: pores

detected: pores at different depths, boundaries between the layers with different densities (carbon deposited layers). Figure 4 shows a B-scan of a part of an isotropic pyrographite sample 10 mm thick with pores at depths of 3 mm to 9.5 mm (indicated by arrows). Figure 5 shows a B-scan of a part of a sample with a boundary between regions with different densities (two deposited layers) (indicated by the dotted arrow).

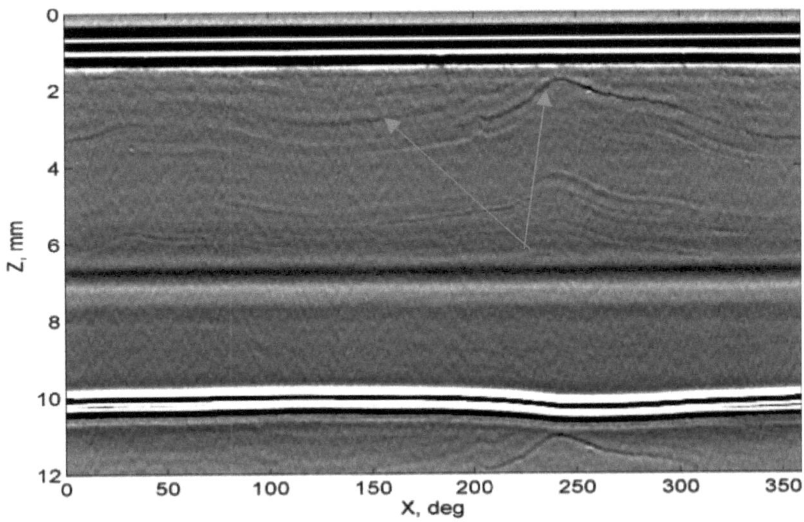

Fig. 5. B-scan of isotropic pyrographite sample: boundary between carbon deposited layers

Some regions exhibited both types of defects (Fig. 6)

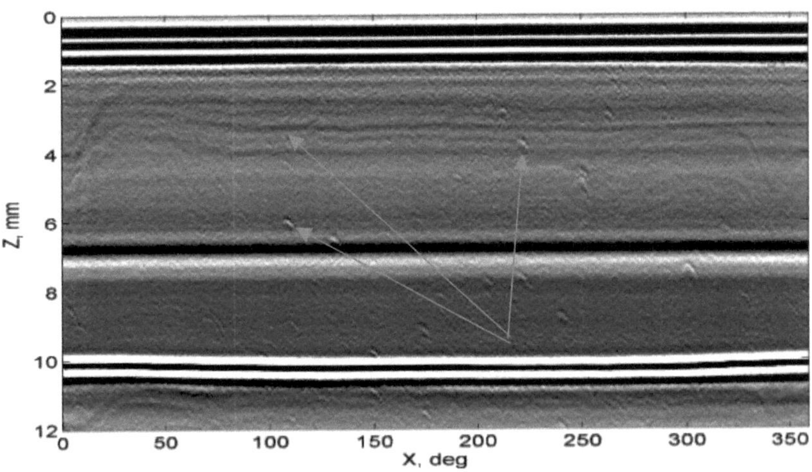

Fig. 6. B-scan of isotropic pyrographite sample: boundary between carbon deposited layers and pores

5 Conclusions

The study shows that broadband ultrasonic spectroscopy is able to detect such defects in isotropic pyrographite as pores and boundaries between carbon layers of different density. This ability of ultrasonic spectroscopy can be used to specify the parameters of the process of manufacturing isotropic pyrolytic graphite by chemical vapor deposition, as well as to study changes in its structure under different influences.

The nondestructive testing method using broadband ultrasonic structuroscopy ensures the detection of pores 0.3 mm and more in size at a depth of 16 mm and more.

Acknowledgments. This work was carried out with financial support from the Ministry of Education and Science of the Russian Federation in the framework of the Increase Competitiveness Program of NUST ‘MISiS’ (no. K1-2015-025).

References

1. Pierson, H.O.: Handbook of Carbon, Graphite, Diamond and Fullerenes. Noyes Publication, Park Ridge (1993)
2. Choe, J., Lim, J.W., Kim, M., Kim, J., Lee, D.G.: Durability of graphite coated carbon composite bipolar plates for vanadium redox flow batteries. Compos. Struct. **134**, 106–113 (2015)
3. Pageot, J., Rouzaud, J.N., Ahmad, M.A., Deldicque, D., Gadiou, R., Dentzer, J., et al.: Milled graphite as a pertinent analogue of French UNGG reactor graphite waste for a CO2 gasification-based treatment. Carbon **86**, 174–187 (2015)
4. Leiro, J.A., Heinonen, M.H., Laiho, T., Batirev, I.G.: Core-level XPS spectra of fullerene, highly oriented pyrolitic graphite, and glassy carbon. J. Electr. Spectrosc. Relat. Phenom. **128**, 205–213 (2003)
5. Maslova, O.A., Ammar, M.-R., Guimbretiere, G., Rouzaud, J.-N., Simon, P.: Determination of crystallite size in polished graphitized carbon by Raman spectroscopy. Phys. Rev. B **86**, 134205 (2012)
6. Salver-Disma, F., Tarascon, J.-M., Clinard, C., Rouzaud, J.-N.: Transmission electron microscopy studies on carbon materials prepared by mechanical milling. Carbon **37**, 1941–1959 (1999)
7. Wirth, R.: Focused ion beam (FIB) combined with SEM and TEM: advanced analytical tools for studies of chemical composition, microstructure and crystal structure in geomaterials on a nanometer scale, Chem. Geol. 261 (2009) 217. Chem. Geol. **261**, 217229 (2009)
8. Le Guillou, M., Rouzaud, J.N., Deldicque, D., Toulhoat, N., Pipon, Y., Moncoffre, N.: Structural and nanostructural behavior of deuterium implanted Highly Ordered Pyrolytic Graphite investigated by combined High Resolution Transmission Electron Microscopy Scanning Electron Microscopy and Raman microspectrometry. Carbon **94**, 277–284 (2015)

Assessment of Protective Structures on the Basis of Their Acceleration Caused by Blast Wave

Jiří Štoller(✉) and Branislav Dubec

Faculty of Military Technology, Department of Engineer Technology, University of Defence, Kounicova 65, 662 10 Brno, Czech Republic
{jiri.stoller, branislav.dubec}@unob.cz

Abstract. The purpose this study was to assess the protective structures of different shapes on the basis of their acceleration caused by blast wave. Six variants of models was assessed in the software environment Ansys Autodyn, an explicit tool for modelling the nonlinear dynamics of solids, fluids, gases and their interactions, where the compared parameter was the value of the acceleration of the structures loaded with blast wave. This study represent the second step in the search for optimal shape of protective structure with regard to limitations such as building materials on the base of concrete, low manufacturing cost, modularity and easy transportability and follows the conclusions from the study [11].

Keywords: Acceleration · Ansys Autodyn · Protective structures · Shape optimization · Blast wave · FEM

1 Introduction

The changes in the security environment and development of armed conflict receding the importance of classical field fortification. The reason to that is a change in the nature of conflicts, tactics, equipment of troops, where the usage of small, highly mobile units comes to the foreground. While the danger of massive conflicts is still present, in present day, the main focus of armed troops is directed towards local conflicts that can very quickly change its center of gravity [5].

Another aspect of the changing nature of conducting operations is the deployment of armed forces operations in diverse and unpredictable geopolitical environments. The possible threats, which arise from the asymmetrical nature of conflicts, place demands for research and development of new ways of force protection. While in combat operations, the task is to minimalize life losses of troops connected with direct combat, there is an equal effort to eliminate the losses in the context of other tasks. The army personnel can support diplomatic efforts, cooperate on mitigation of consequence of humanitarian problems in unsteady environment, support civilian reconstruction teams etc. Huge defensive positions, fortifications and places of gathering were thus replaced by smaller military bases and peacekeeping camps, where also the non-military personnel are present. Besides the attacks performed by conventional weapons, the main

A. Kravcov et al. (eds.), *Durability of Critical Infrastructure, Monitoring and Testing*, Lecture Notes in Mechanical Engineering, DOI 10.1007/978-981-10-3247-9_2

threat to camps and bases comes from the usage of improvised explosive devices and charges carried by persons, luggage or vehicles [12].

Despite the relatively large amounts of resources of the Army of Czech republic dealing with designing protective structures in both, the theoretical and practical aspects in, there has not been yet introduced to the usage a protective structure, that would meet the all the requirements, which contemporary force protection requires. Contemporary protective structures are characterized by one or more deficiencies that reduce their protective value. The most relevant is the lack of optimization [9], and therefore its low resistance to the effects of the blast waves. So far, this deficiency was mostly ad hoc eliminated by the usage of adequate amount of construction material of outer protective layers of the structures [8]. Another drawback of protective structures is a relative time-consuming and labor-intensive construction. The structure is usually made of several different materials and different parts, which require heavy machinery for their assembly which is also connected to another major drawback and that is demanding character of their transportation [4].

This study continues in the search of a new protective structure that will remove deficiencies mentioned in previous paragraphs and open the next phase that is the creation of its structural design.

2 Acceleration

Blast waves cause two types of acceleration of objects, such as structures, vehicles and biological specimens. The first is characterized by its very high value, accompanied by small values of translation of objects. In general, its presence is associated exclusively with the phenomena of explosive nature. The second one is characterized by lower value, where higher values of translations of object are present. This acceleration is typical also for other phenomena like traffic accidents [3].

In the context of force protection, protection of materials or equipment to the effects of explosions are these two types of acceleration counted among the devastating parameters that must be eliminated or reduced to acceptable levels. Among injuries caused by the effects of acceleration are mainly first limb injuries, spinal or brain ranging from nerve damage, through soft tissue injury, fracture or shatter bone, crushed cartilage, ligament rupture etc. All these injured arise due to the passage of the shock wave and the body's inability to absorb this energy. Effects of the second type causing injuries and damage due to different acceleration of parts or the secondary collision of these parts with other objects [7].

Acceleration of the construction caused by the blast wave of the explosion that is transmitted to human beings located in structure, is therefore appropriate indicator for assessing the level of ballistic resistance of protective structure. Since the objective of this study was to find a suitable shape of protective structures, it was not necessary to find out the real value of the acceleration of the structure and evaluate its effects on the personnel. The assessment of the quality of ballistic resistance of the designed protective structure of different shapes to the effect of the blast wave is obtained by comparing values of acceleration of individual models of these structures.

3 Ansys Autodyn Simulation

Ansys Autodyn is explicit analysis tool for modeling the nonlinear dynamics of solids, fluids, gases and their interactions. When calculating the propagation and effects of the pressure wave from the explosion as a physical problem, its mathematical model based on a system of partial differential equations is discretized and afterwards solved using numerical methods. Ansys Autodyn represent a standalone executable component of an integrated software environment Ansys Workbench (AW) where the entire process of the creation of the simulation can be managed within a few subsequent steps.

At the beginning of the creation of a computational model, explicit dynamics component was chosen, where the tasks connected with explosions can be solved. Afterwards, the proper materials used in the simulation had to be chosen. Material Library of AW is divided into several categories according to the nature and complexity of the simulation material models, where explicit material model for concrete named CONC140Mpa, representing the building material of the protective structure was chosen.

The next preparatory step consist of creation of the models geometry that can be used in the simulations. Since in our case our task was to simulate the effect of an explosion on a protective structure, the base geometry represented by the model of the protective structure created by CAD software Autodesk Inventor was imported into an software component of AW for creating the geometry for simulation called DesignModeler (Fig. 1). To reduce the computational demands, model was adapted to the shell structure with a volume of 0.0046 m^3. After that step, it was proceed to a creation of a mesh.

Meshing is a process of a creation of the numerical network of the model geometry. Through this process the model is divided into a finite number of elements, where each element carries information about the detected parameters in its node (Fig. 2). Network quality greatly affects the accuracy, speed and fluency of calculation and therefore a refinement and adjustment of a default network with regards to the size of the minimum element and the overall quality of elements was made. Because of the necessity to enter at least one initial conditions, the model was defined with the direction of the

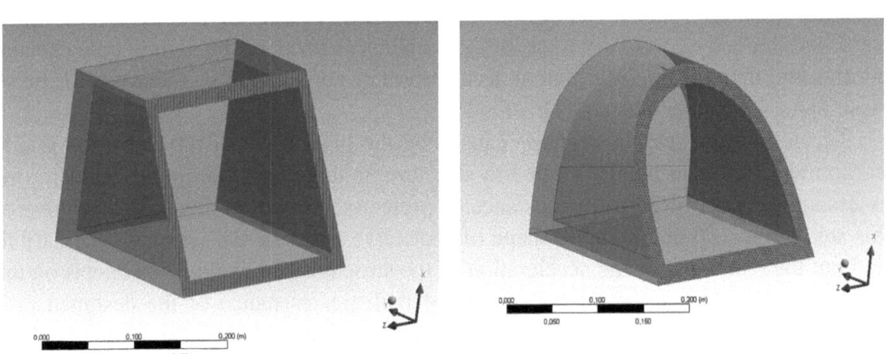

Fig. 1. Cross-sections of the models in DesignModeler

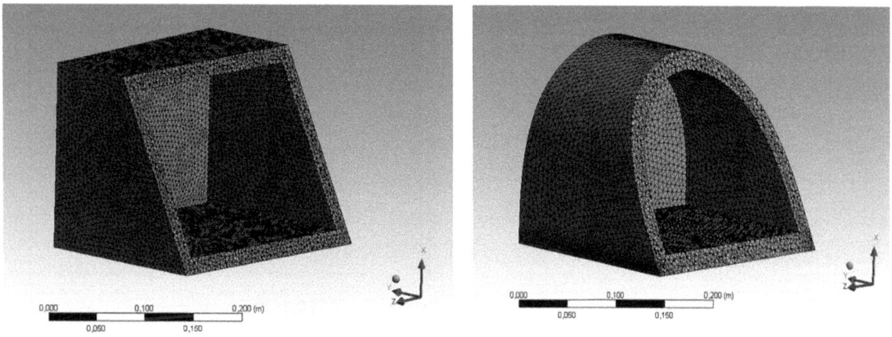

Fig. 2. Mesh of the models

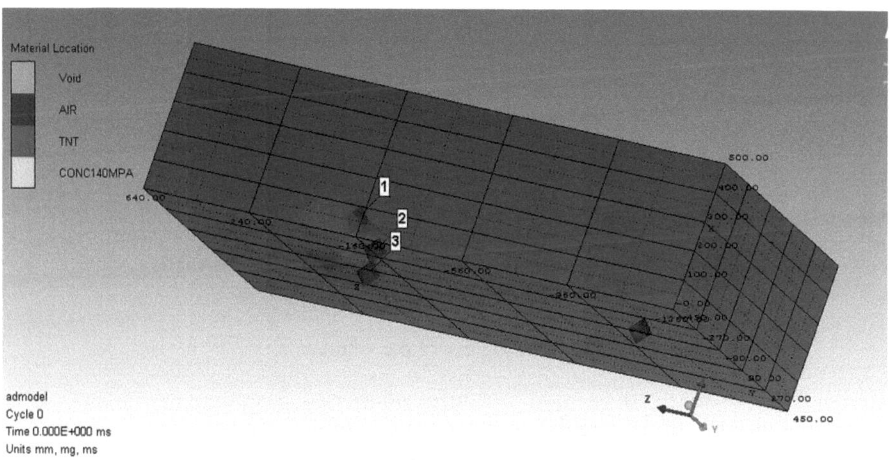

Fig. 3. Computational domain

gravitational acceleration with the direction of negative x. Pre-prepared model was subsequently linked and opened in Ansys Autodyn, where another settings was specified and where the final calculation of the propagation of blast wave and its effect on the protective structure was simulated.

Besides the already chosen material model of the construction, another two materials required for the interaction between a physical model of an explosion, such as air and TNT were loaded. Air was used as a filling of the computational domain of propagation of blast waves and TNT filled part of this domain, representing the explosive charge of total weight 106.4 g. The detonation point was situated to a distance of 1 m from the front edge of the building (Fig. 3) in the center of the placed charge. On the five planes of the domain except for bottom plane YZ of the domain representing the earth's surface was applied Flow_Out condition, which allowed free movement of material through this areas.

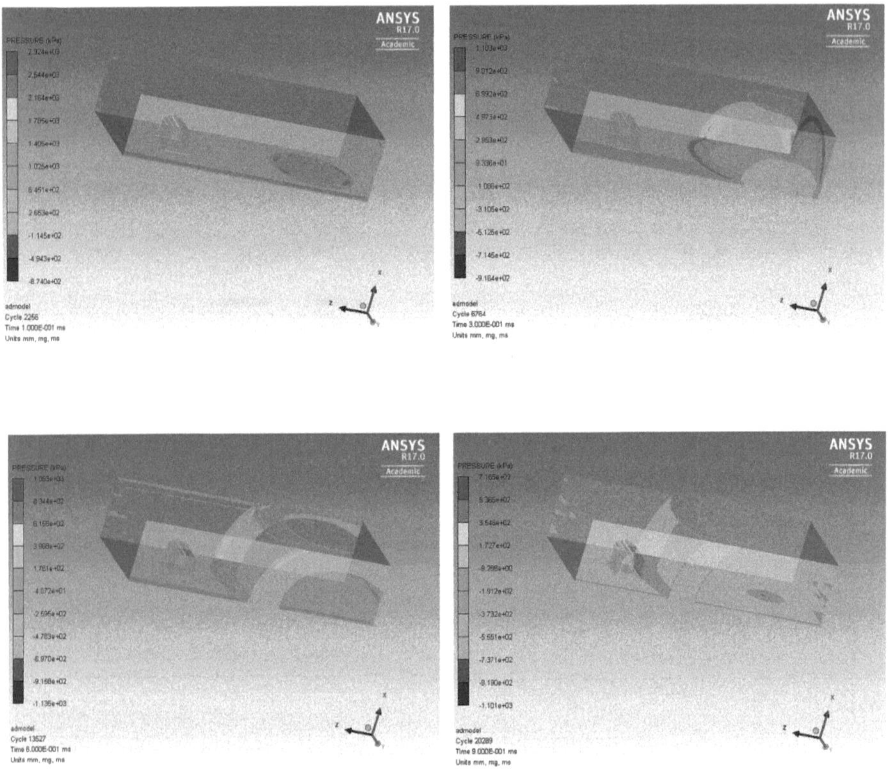

Fig. 4. Progress of the simulation

Before the start of the simulation, three gauge points for recording the value of the acceleration in direction of axis negative z were placed into the shell of the model. After this step, the duration of simulation was set and calculation was launched for the first model (Fig. 4).

4 Variants Assessment

Overall six variants of protective structures was assessed (Fig. 4). These variants of protective structures represent the entire range of values of the drag coefficient obtained by CFD conducted in paper [11] and the both factors, requirements for the dimensions and volume of internal space [4] of protective structure or the possibility to construct them were taken into account by their selection. The model number 1 (Fig. 5) represent the specimen with the lowest value of drag coefficient, the model number 6 represents the model with the highest value (Fig. 5).

Table 1 shows the values of the parameters of models in ascending order according to their drag coefficients [11], as well as absolute values of the maximum values of their

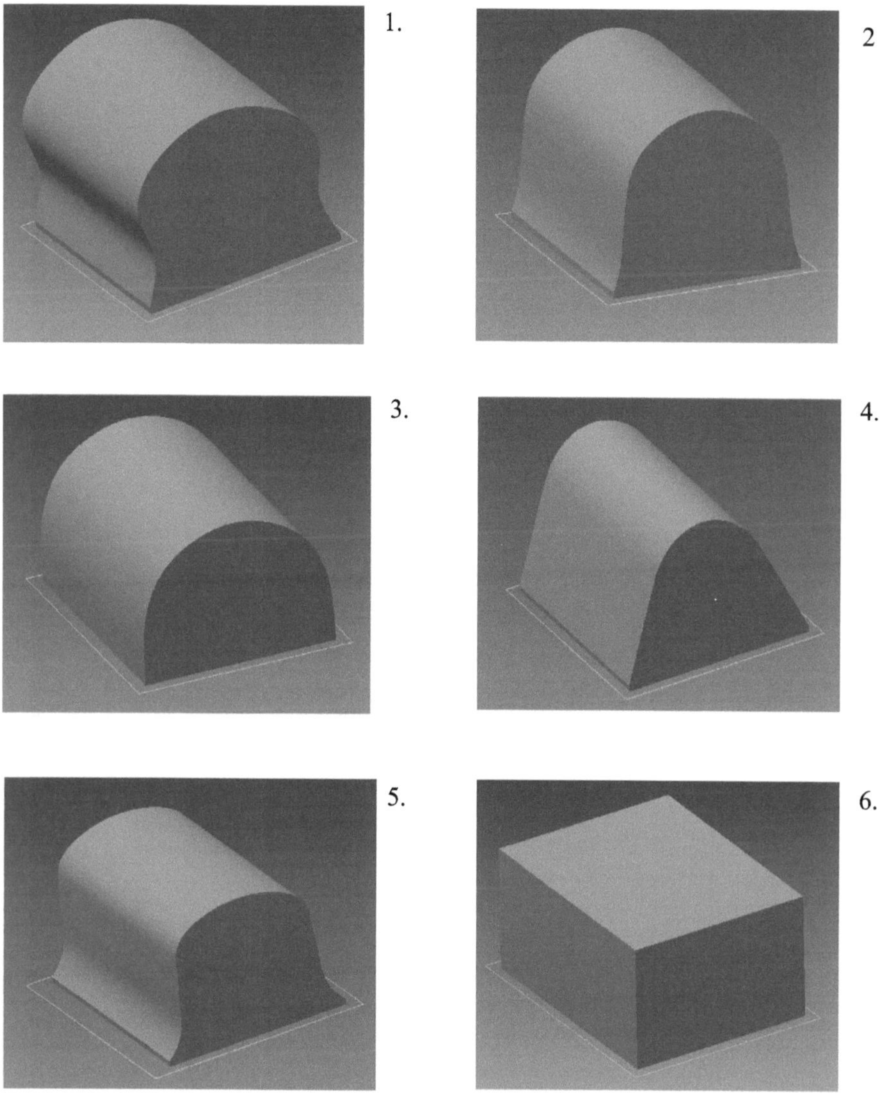

Fig. 5. Shapes of assessed models

acceleration obtained from three gauge points placed on the models caused by blast load obtained from the simulation by the usage of Ansys Autodyn.

The bar graph (Fig. 6) is a visualization of the data from Table 1. It is evident, that the correlation among the value of drag coefficients and the value of acceleration is not present. However from the assessments by the usage of both, CFD software environment Anys Afluent and Ansys Autodyn, it is clear, that the traditional shape of the protective structure (model number 6) is not optimal from the ability to resist the effects of loads from blast waves point of view.

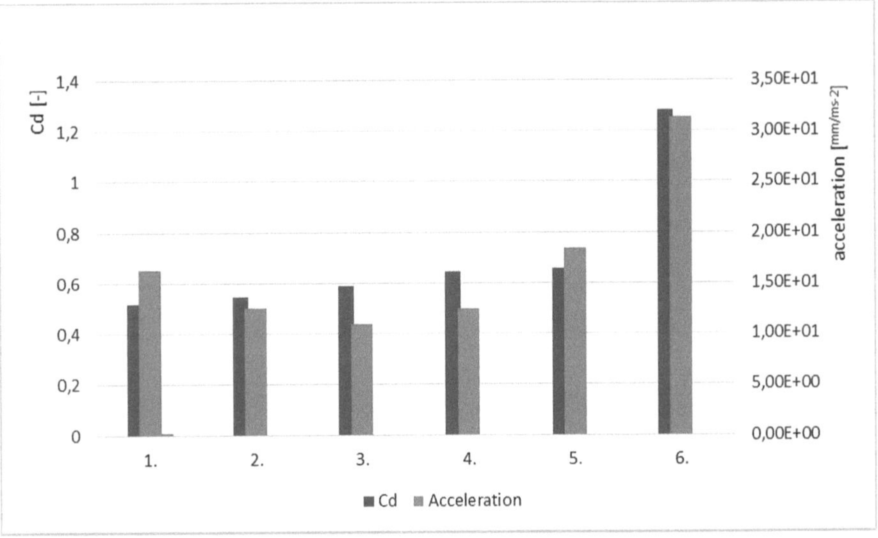

Fig. 6. Drag coefficient and acceleration of models

Table 1. Input parameter values and their range

Model num	Parameter DS_1 [mm]	Parameter DS_2 [mm]	Parameter DS_3 [mm]	Parameter DS_4 [mm]	Parameter DS_5 [mm]	Force – z [N]	Drag coeff C_d	Accel [mm/s.10^{-6}]
1.	88,3	119,4	112,6	95,1	112,5	44,67	0,5186	16,3
2	83,5	101	105	107	110,7	47,12	0,5471	12,5
3.	80	105	117	120	120	45,53	0,59	10,9
4.	60	73	91	100	112	55,61	0,645	12,4
5.	98	87,9	97,1	94,3	100,7	56,59	0,657	18,4
6.	Cuboid, A × B, A = 240 mm, B = 190,7 mm					110,1	1,282	31,3

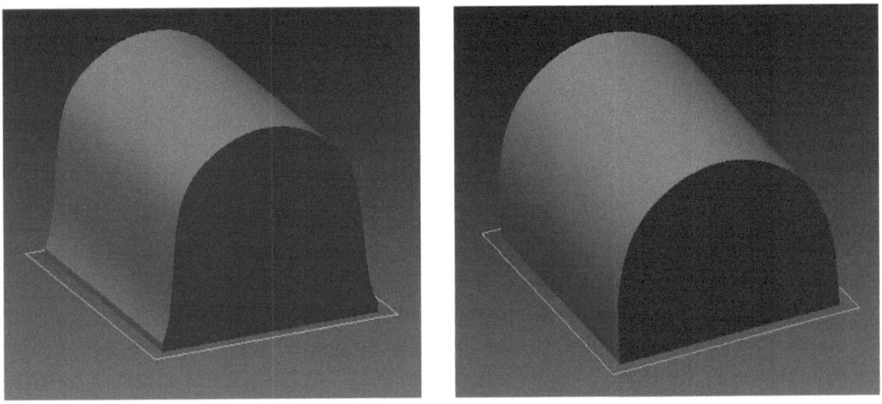

Fig. 7. Expedient shapes of protective structure

This assessment also led to conclusion to exclude model number1 and model number 5. Three models felt into consideration, model number 1, model 2 and model 3. After excluding model 4 because of its unfavorable internal topology, model number 2 and model number 3 became models (Fig. 7), which structural design will be worked up.

5 Summary

This work presents the second step in design of new protective structure to the effects of the blast wave and follows the results obtained in previous study, where the numerical simulation in the software environment Ansys Fluent based on computing model of the external flow around the models [1] of protective structure was conducted as a method for finding the optimal shape. The criterion for comparison was represented by the value of drag coefficient [11].

In this study, six models of different shape were assessed in the software environment Ansys Autodyn to the effect of the blast created by surface explosion, where the value of acceleration of the construction caused by the blast wave from the explosion was compared. Based on the results of both simulations the most expedient shapes of protective structures optimized to the effect of explosive blast wave were selected.

The creation of the structural design is the task of the third step in designing the new protective structure. To reduce production costs, reduce transport costs and demands and to facilitate assembly of the structure, which were the original requirements and conditions of the new design, it seems expedient to divide the overall construction of the protective structures into several structural prefabricated elements. This design can fulfil the requirements placed on protective structure, as well as demands for its resistance against the means of destruction of the enemy [10]. The basis of the structure can be created by several concrete frames as this type of structural design was used in the past for construction of protective structures. These were mainly soil covered shelters, outposts, hideouts, observation posts etc. [6]. Since the advantages of frame-type structures is their considerable resistance, durability and relative ease of production as well as vast means for their transport, the frame is a structural element suitable for construction of protective structures in present day.

The possibilities for further advancements in development of protective structures can be divided into several categories. Ansys Autodyn as a component of an integrated platform, Ansys Workbench, along with other components of the platform offers extensive options for dealing with the design and evaluation of protective structures. One of the main conditions to achieve accurate results is the necessity to input correct parameters [2]. The material models used in this simulation did not correspond to the planned building material model of the new protective structure. Although the nature of the simulation did not require its existence, its creation and incorporation into the material library of the software would help improve the accuracy of the computational model and the results obtained in subsequent simulations.

Another improvement of the quality of protective structure is the development of manufacturing capabilities allowing the production of shaped elements, which removes some of the constraints imposed on the geometry of the protective structure.

The creation of a rotationally symmetrical body with the infinite planes of symmetry could help to further minimize the interaction of blast waves with a protective structure [9]. Possibility of production of shape and dimensionally more diverse components of the structures would allow to emergence a wider range of shape and dimension solutions of protective structures.

The design of the new protective structure presented in this study will enable units operating in foreign missions increase their level of protection. The elimination of threats which are the units in this type of operations exposed to requires new solutions and continuous improvements in force protection.

Acknowledgment. The work presented in this article has been supported by the Czech Republic Ministry of Defence - K 201 Department development program "Development of technologies in weapons construction, ammunition, weapons instrumentation, materials engineering and military infrastructure".

References

1. ANSYS, Inc. ANSYS Fluent Theory Guide, Release 15, USA, SAS IP, Inc., November 2013
2. ANSYS, Inc. Design Exploration, Release 12.1, USA, SAS IP, Inc., November 2009
3. Boyd, S.D.: Acceleration of a plate subject to explosive blast loading - trial results, Australia, Department of Defence, Maritime Platforms Division Aeronautical and Maritime Research Laboratory, March 2000
4. Cerny, F., Porocak, L.: Fortification II. S-3151/2. Military academy of ground forces, Vyskov (1999)
5. Coufal, D.: Modelling of effects of blast wave on the protective structures. Dissertation Thesis. University of defence in Brno, Faculty of military technology, Brno, 102 p. (2013)
6. Coufal, J., Kaplan, V.: Protective structure. S-297/III. Military academy in Brno (1993)
7. Courtney, M.W., Courtney A.C.: Working towards exposure thresholds for blast-induced traumatic brain injury: thoracic and acceleration mechanisms, USA, U.S. Air Force Academy, Force Protection Industries
8. Field fortification: Surface and Soil-covered Shelters. Zen-2-1/1 Prague: MNO (1972)
9. Gebbeken, N., Doge, T.: Explosion protection-architectural design, urban planning and landscape planning. Int. J. Protective Struct. **1**, 1–22 (2010)
10. STANAG 2280 - MC ENGR (Edition 1) (Ratification draft 1) – Design Threat Levels and Handover Procedures for Temporary Protective Structures. NATO Standardization Agency, June 2007
11. Stoller, J., Dubec, B.: Design and assessment of shape of protective structure by usage of CFD software environment Ansys Fluent. Manuscript submitted for publication (2016)
12. Zalesky, J.: Military engineering support of force protection in foreign missions. Dissertation thesis. University of defence in Brno, Faculty of economy, Brno, 162 p. (2012)

The Possible Means Suggested for Improvement of Evaluation of Low Endurable Terrain

Klára Cibulova(✉), Matouš Formanek, and Martin Priesner

Department of Engineer Technology, University of Defence, Kounicova 65, 66210 Brno, Czech Republic
klara.cibulova@unob.cz, {mathew,m.priesner}@seznam.cz

Abstract. This article deals with the improvement of means for evaluation the trafficability of terrain in the frame of the Engineer Corps of the Czech Army. There are described the most used instruments the telescopic penetrometer PT-45 used in the Czech Army and the cone penetrometer from the Trafficability Test Set used by some NATO countries. There is written about problems the Engineer Corps has to face. It means that there are not enough instruments. The second weakness for the proper evaluation of the trafficability of terrain is the lack of materials for working with these instruments. So the steps making for removing these problems described here.

Keywords: Trafficability of terrain · Telescopic penetrometer PT-45 · Trafficability test set · Cone penetrometer · Vehicle cone index · Rating cone index

1 Introduction

Extraordinary situations come about time to time in our region, as well as in other regions in the world. These situations are usually brought about by different natural influences. Due to these situations some parts of communications became no passable (the bridge damage by the flood etc.). Then there is need to transport to an unapproachable area in the shortest period of time. The detour is not always possible or is too long. Then you have to go through the terrain. The question is, if the terrain is suitable for coming through for given type of vehicle or not. Different types of instruments are used for judging the trafficability of terrain.

The terrain is capable of carrying the load caused by moving vehicles due to his shear strength. In practice, the determination of shear strength using penetrometer which may be static or dynamic. Penetrometer is a tool used to diagnose bearing capacity of the soil. It can, particularly in construction, agriculture and forestry, where the findings, we determine the required degree of compaction surface. Penetrometer measuring principle is based on the strength that is needed to tip penetrometer sinking into certain depths. Based on the results of measurements can be determined for example passability of terrain. To achieve more accurate results, it should be done on the desktop always more measurements. Some penetrometers allow measurement of

A. Kravcov et al. (eds.), *Durability of Critical Infrastructure, Monitoring and Testing*, Lecture Notes in Mechanical Engineering, DOI 10.1007/978-981-10-3247-9_3

the amount of moisture in the soil, which is used among other things for measurements on other planets. There are many types penetrometer, are usually round or conical shape [1].

As it was mentioned above in the Czech Army there are used two instruments. The first one is the telescopic penetrometer and the second is the cone penetrometer from the Trafficability test set. The procedure of working with the telescopic instrument is described in the Field Manual Military Roads and Ways [2]. Working with the test set is described in the Field Manual 5-430-00-1 Planning and Design of Roads, Airfields, and Heliports in the Theater of Operations — Road Design [3].

Both instruments and working with them are described here as well as the suggested ways of improvement and simplifying of evaluation for the solders.

2 Measuring with Penetrometers

Telescopic Penetrometer PT-45. It is aged device, which is easy to operate and its major strength is fast measurement Fig. 1.

Fig. 1. Telescopic penetrometer PT–45 [2].

The telescopic penetrometer is equipped with a thorn Fig. 1. When the penetrometer is pushed into the ground the thorn penetrates the soil and the pressure used to pierce the terrain is displayed on the side scale of the device. The amount of pressure depends on the depth of penetration and type of the soil. The units on the scale are in MPa. The depths in which the pressure is read are 5, 10, 15, 20, 25 and 30 cm on the same place. Each measurement is made three times in one-meter distance.

The trafficability (number of vehicles that could pass through the area) is determined from the evaluation report. The data that has been read from the scale are written to this report. All values from same depth are averaged. In the next step first free values are averaged and same happen with the rest. From the lower number on the evaluation table it is found out the number of vehicles that could pass the terrain according to their total weight. If the terrain has high altitude the number of vehicles must be adjust.

As a low supporting terrain we consider the terrain where the telescopic penetrometer measures the value lower than 3 MPa [2] (Fig. 2).

The Trafficability Test Set. Another possible instrument used for the measuring the trafficability is the cone penetrometer Fig. 3. It is one component of the Trafficability test set. This set contains canvas carrying case, one cone penetrometer, one soil sampler, remolding equipment and a bag of hand tools (spanners, screwdrivers, wrenches etc.). This penetrometer is more difficult to measure but on the other hand its outcome is much more accurate.

Fig. 2. Measurement with the telescopic penetrometer.

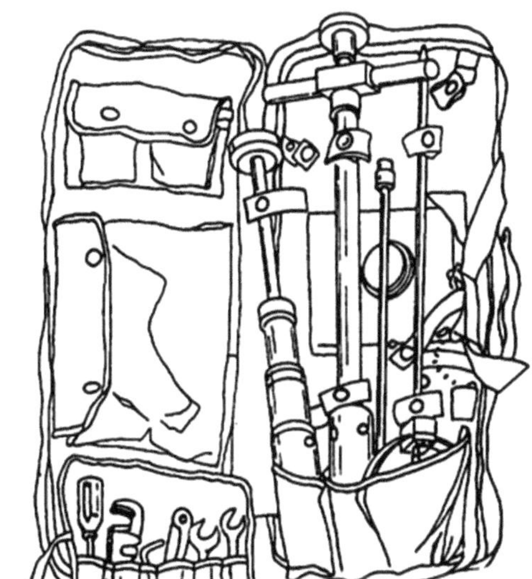

Fig. 3. The trafficability test set [3].

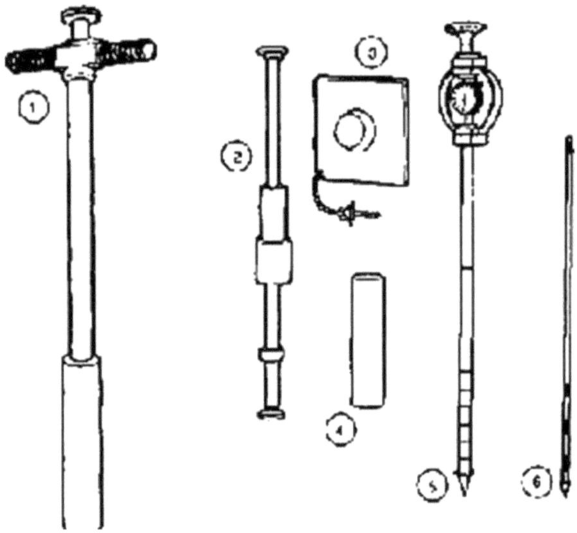

Fig. 4. The remolding test set [3].

The cone penetrometer is used for determining the resistance against permeating of a thorn in low-strength soils. The dynamic penetrometer exist as well and it is used for high-strength soils. When the penetrometer is pressed to the soil, the ring shows the needed value for pushing the thorn to the soil. The cone penetrometer could not be used for the gravels. Gravels are considered to be trafficable for 50 passes and any problem can be search out by the reconnaissance.

Whether the terrain is trafficable or not is found out from the comparison between the index of soil and the index of vehicle. The index of soil is called the rating cone index RCI and the index of vehicle is called the vehicle cone index VCI.

The soil index is product of two indexes — cone index and remolding index. Both these indexes are set with the soil-trafficability test set. The assessment of the cone index CI is the resistance of the soil in its unbroken condition. The penetrometer is pressed to the soil and the ring shows you the power needed for the pressing. This power is the index of the resistance and it is called the CI [3].

While the vehicles are passing the terrain the features of the soil are changing. There is the remolding index which describes the changes in the soil during the passes. To determine the RI we need the soil sampler and the remolding test Fig. 4. We take the soil in the unbroken condition to the soil sampler and measure the CI with the cone penetrometer. Then we simulate the changes during passes of vehicles. This is made by the weight which is let down to the soil. The number of falls depends on the type of soil. Then we measure the CI in the changed soil again. The rate between the CI before the simulate changes and after them is the remolding index RI. The product of CI and RI is the RCI – soil index.

The vehicle cone index VCI is a value given to the vehicle for given number of passes — usually one or fifty. There is opportunity to estimate the VCI for the vehicles

Fig. 5. Measurement with the cone penetrometer.

when we do not know the index. We have to know the features of the vehicle such as — the weigh, type of motor, clearance, number of wheels etc. Then we can calculate the VCI. If the VCI is higher than the RCI it means that the vehicle index is higher than the soil index and the given number of vehicle can go through the given area. Whether it is conversely, RCI is higher than VCI; the vehicles are going to sing into the soil (Fig. 5).

For conventional types of vehicles used in some NATO countries, the values of vehicle cone indexes are known and tabulated. On the other hand, we have to count the VCI for the vehicles used in the Army of the Czech Republic. First we are able to count the mobility index MI and after that we can find the vehicle cone index from a curve in a chart for one or fifty passes. Here is the formula for counting MI [3] (Fig. 6):

$$MI = \left[\frac{(CPF) \times (WF)}{(TF) \times (GF)} + (BF) - (CF)\right] \times (EF) \times (TSF), \tag{1}$$

where:

PF - contact pressure factor,
WF - weight factor,
TF - track factor,
GF - grouser factor,
BF - bogie factor,
CF - clearance factor,
EF - engine factor,
TSF - transmission factor.

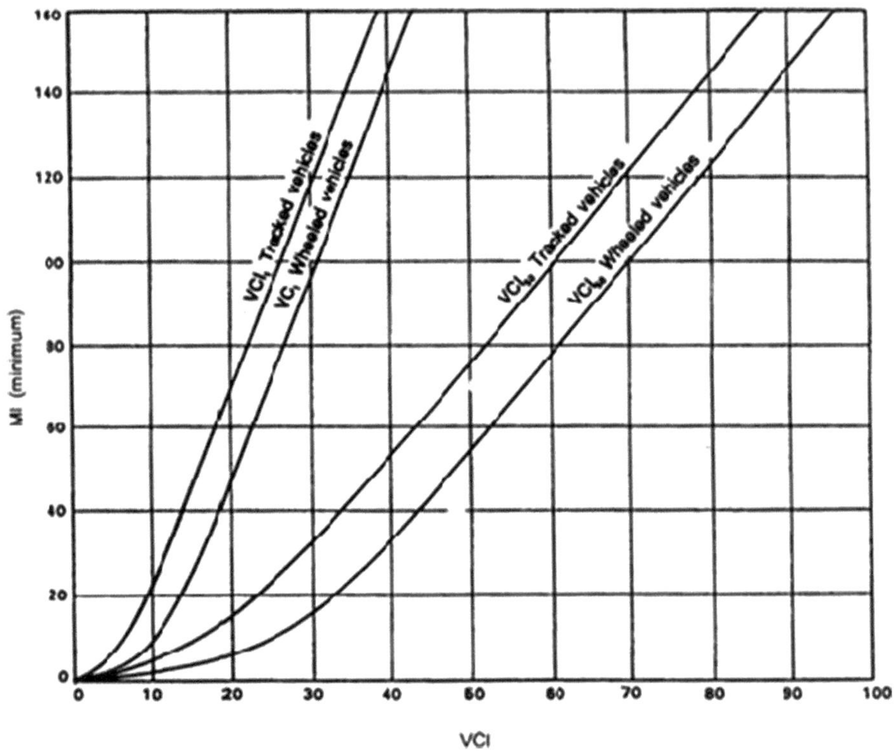

Fig. 6. Transfer of mobility index to vehicle cone index [3].

3 The Suggested Proposal of Simplifying the Evaluation of Trafficability

Questionnaire. For a better understanding of the problems with evaluation the trafficability in the ACR was created questionnaire consisting of fifteen questions. Questions were focused on usability and work with the penetrometer presence of all necessary materials, guidelines and evaluation procedures for individual species. Engineer battalions as well as training facility in Vyskov come into contact with penetrometers. The respondents were members of the Engineer Corps and members of the Military Academy, who are responsible for the implementation of basic, vocational and special training, career military education, training of military professionals and active members of the ACR advances in engineering training. They were answered following 15 questions:

1. Do you operate with penetrometers?
2. How often do you work with this instrument?
3. How would you described the working with the penetrometer?
4. How many manuals Žen 2–16 including the annexes do you have available?
5. Are the penetrometer you have functional?

6. Are the penetrometers in a good conditions?
7. Is there something you would like to change on the penetrometer?
8. Do you face some problems with the penetrometer?
9. How do you store the penetrometers?
10. How many new pieces of the penetrometer would you appreciate?
11. Would you be interested in a new penetrometers?
12. How do you evaluate the measured values?
13. Have you met with other kinds of penetrometers?
14. Do you visited Information Portal of Corps of Engineers (www.ipzv.unob.cz)?
15. Would it help you if they are evaluating tables online for download on the websites?

Questionnaire shows that there are following problems with evaluation the trafficability of terrain. The first one is that there are different approaches to the evaluation. They have some tables, but no procedures. The second problem is the lack of instruments. So due to the questionnaire they would really appreciate the opportunity to download the procedures and tables from the Engineer portal as well as the possible way to have it in electronic way. And of course some other instruments would be a great help for them.

So because the soldiers has to deal with incomplete or missing documentation to the above mentioned instruments, it has been started creation of a guide for evaluating the value of field measurement throughput terrain.

Protokol průchodnosti terénu pro PT-45

Vypracoval	Datum a čas	Číslo protokolu	Místo
Momentální teplota .... °C	Počasí:	Srážky:	Terén:

Nákres prostoru	Poznámky

Vzdálenost měřeného místa (m)	Hloubka vpichu	Zjištěná penetrační únosnost			Průměr jednotlivých vrstev	Průměr z průměrů tří vrstev	Minimální průměr	Počet průjezdů vozidel o celkové hmotnosti			Nerovnost terénu v cm/m² v úseku na 10 m		Snížená průjezdnost vozidel o celkové hmotnosti			Porost	Hornina
		1.	2.	3.				4,5 t	9 t	15 t			4,5 t	9 t	15 t		
	5										5	x 0,88					
	10										10	x 0,75					
	15										15	x 0,6					
	20										20	x 0,4					
	25										25	x 0,1					
	30										30<	1 kolej = 1 vozidlo					

Fig. 7. The suggested protocol for measurements with the telescopic penetrometer PT-45.

Vypracoval:	Martin Priesner	Datum a čas	23.3.2016 8.30	Teplota:	13
Počasí:	oblačno, přeháňky	Místo měření	Vyškov, Hanácká louka	Poznámky:	

Vzdálenost měřeného místa	Hloubka vpichu	Zjištěná penetrační únosnost			Průměr údajů jednotlivých vrstev	Průměr z průměrů tří vrstev	Minimální průměr
		1.	2.	3.			
180	5	2,0	4,0	3,0	3,0		
	10	4,2	5,0	2,5	3,9	3,6	
	15	4,0	4,0	3,3	3,8		3,6
	20	2,2	5,0	4,0	3,7		
	25	4,0	3,9	4,2	4,0	3,7	
	30	4,0	3,0	2,9	3,3		

Nerovnost terénu v cm/m2 na úseku 10 m	
	13

Počet průjezdů vozidel o celkové hmotnosti		
15 t	9 t	4,5 t
4	10	25

Snížená průjezdnost vozidel o celkové hmotnosti		
15 t	9 t	4,5 t
3	8	19

Fig. 8. The suggested protocol for measurements with the telescopic penetrometer PT-45.

In the first step is going to be the theoretical guidance on the measurement. This means writing procedures for measurement - selecting a suitable place to measure the correct handling of the device atc. The second partial objective is to establish protocols for recording the measured values during measurement. These protocols will be available in both printed and electronic form (Fig. 7).

The result is a complete tool for evaluating the trafficability of terrain. The manual will be used during field measurements, in the exercises or in crisis situations. Procedures, protocols and evaluation forms will therefore be placed on the information portal Corps of Engineers for their further utilization. And the more there will be made videos with procedures of working and evaluation for easier understanding.

All this has been creating by the students of the Department of Engineer Technology of the University of Defence as their student's research work. And now there is working on the second part – the guide or manual for the second instrument – the cone penetrometer (Fig. 8).

4 Reverse Engineering of the Telescopic Penetrometer

Drawing Functional Sample. Examining the subject in order to determine the principle of its operation and in order to make a well-functioning object is engaged in reverse engineering. For this reason, the drawing was made by penetrometer PT-45 connector. Part of the drawing is a 3D model and animation made in design program Autodesk Inventror (Fig. 9).

Animation demonstrates the functionality and principle of penetrometer is shown on Fig. 10.

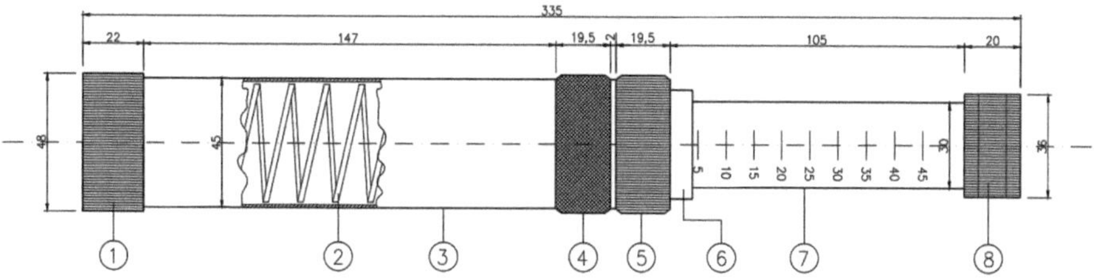

Fig. 9. Drawing of penetrometer PT-45.

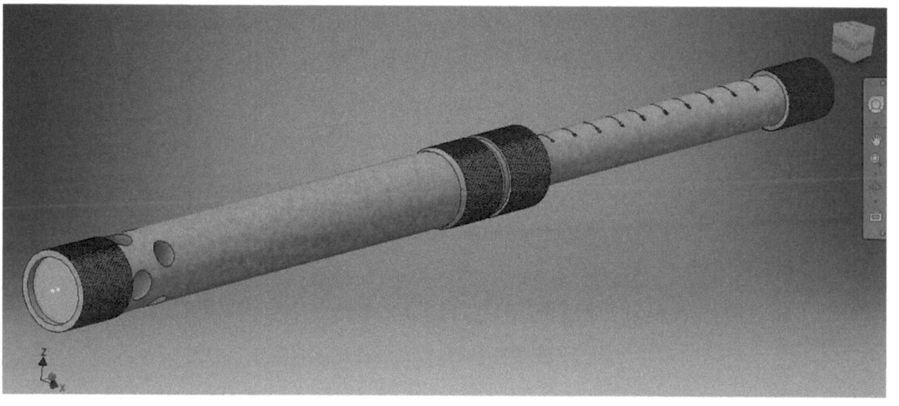

Fig. 10. 3D model of penetrometer PT-45.

5 Summary

Finally it is necessary to emphasize the importance of knowledge the trafficability of a terrain. There are two spheres of utilization — military and civil. In the military sphere it is all about very quick negotiation of a terrain during military actions. The way through terrain could be the shortest way or elected detour due to impossibility of using the original route. In the civil sphere it is usually about the coming through the terrain because of damage of a communication due to some natural influences. Then there is need to get to these weatherworn areas to ensure food, material, reconnaissance of harms. And we cannot forget one of the most important things such as the transport of ambulances and other help. If we are able quickly and reliably judge the trafficability of terrain, we can help in all these mentioned cases above.

On the base of the questionnaire the Department of Engineer Technology tries to improve the evaluation of trafficability of terrain. This means to complete the manuals for measuring and to put them to the Engineer Corps Portal to be available. The videos with procedures of working and evaluation for easier understanding will be there as well. The second part is the reverse engineering of the telescopic instrument PT-45. These materials could be used for future production of new penetrometers.

Acknowledgment. The work presented in this article has been supported by the Czech Republic Ministry of Defence - K 201 Department development program "Development of technologies in weapons construction, ammunition, weapons instrumentation, materials engineering and military infrastructure".

References

1. Cibulová, K.: Diagnostic of Terrain Trafficability with Wheeled Vehicles for the Need of Czech Army. Dissertation. Brno (2007)
2. Žen-2-16 Military Roads and Ways. Praha (1987)
3. Field Manual 5-430-00-1 Planning and Design of Roads, Airfields and Heliports (1994)
4. Information on http://www.strojeprozkusebnictvi.cz

Neural Networks in Back Analysis of Tunnels

Pruška Jan(✉)

FCE, Czech Technical University in Prague,
Thákurova 7, Praha 6, 169 26 Prague, Czech Republic
Pruska@fsv.cvut.cz

Abstract. To ensure the best agreement of the numerical model behaviour and the reality the back analysis can be used. At present, the engineers prefer relatively simple back inverse methods, however it does not necessarily lead to the desired results. Between the current methods to perform back analysis of soil parameters the method based on the artificial neural networks is the one which is used in the Czech Republic. After a short introduction the principles of the prediction of the tunnel deformation using multi-layer neural network with back propagation are described. At the end of the paper the practical application of the neural networks in the back analysis is shown.

Keywords: Neural networks · Back analysis · Tunnels

1 Introduction

Unlike most materials in civil engineering, the soils and rocks exhibit unpredictable and various behaviour due to the process of formation of these natural materials. For this reason, geotechnical engineers often use simplifications and analytical or empirical methods for design. In the case of the tunnels we have to describe the behaviour of the rock mass and construction method and numerical methods have to be adopted. The designer must guarantee that the behaviour of the numerical model is as close to the reality as possible. The back analysis is one of the subsidiary methods that assist in this task. Back analyses can be generally divided into two types – a direct and inverse one. The inverse methods make a modification of the input parameters of the calculation according to the results of observations to achieve the closest agreement between the measured values and the values determined by calculation (there are two optimization techniques - the converging and direct one). Direct methods for back analysis are based on minimizing of an error of a function defining the results properties. These methods generally use iterative procedures applied to a nonlinear error function of unknown parameters. The beginning of using the back analysis in geotechnics goes back to the seventies. Annamaria Cividini published one of the first comprehensive work generally focused on using of the back analysis included solved examples [1]. Neural networks application in underground constructions has been developed since 1990s, e.g. Work Lee and Sterling [2] or Moon et al. [3]. Recently the following authors deal with this topic in the world: Yoo and Kim [4], Park [5], Ochmanski [6], Li et al. [7] and Zapletal [8] or Pruška [9] in the Czech Republic.

A. Kravcov et al. (eds.), *Durability of Critical Infrastructure, Monitoring and Testing*,
Lecture Notes in Mechanical Engineering, DOI 10.1007/978-981-10-3247-9_4

2 Artificial Neural Network

Neural Network (ANN - Artificial Neural Network, respectively NN - neural network) is a method based on behaviour of a brain tissue which solves a relationship between a multidimensional input variable and one or more dimensional output variable. The history of NN goes back to the 40 s of the last century when the development of neurological research begun and the basic principles of NN could have been established. In 1958, Rosenblatt defined a mathematical model of the neuron called perceptron as a base of multilayer neural networks [10]. In recent years, NN are a very popular and effective method not only in expert systems but also in independent solving of problems when the conventional algorithmic approach is difficult.

The inspiration for the NN is the structure of a brain tissue of higher organisms, whose knowledge is stored using connections between neurons (cerebral cortex contains 13–15 billion neurons). The body of the neuron (memory cell) is activated by so called dendrites on the base of information signal (it arises internal potential). The information signal generated by an activated neuron's body is transmitted via axon fiber to the neuron output - synapses. The synapses process the signal (regulate the intensity of the neuron response) and afterwards transmit the signal to other neurons (each neuron can be connected with up to 5,000 other neurons).

A mathematical model of a neuron is the basic element of NN (structural and functional) - Fig. 1. It multiplies all input variables x_i by a weight factor w_i. The sum of the weighted input variables $\Sigma w_i x_i$ is then transformed by neuron using activation function f (also called transfer function of a neuron) and becomes the output neuron $y = f\ (\Sigma w_i x_i)$. Activation function expresses the intensity of the neuron response to a given input. The most frequently used activation function is sigmoid $z = 1/(1 + e^{-z})$. The other activation functions can be e.g. linear function, step function, hyperbolic tangent etc. The sigmoid is the most often used one because it is similar to the biological response function.

The weight functions w_i represent the intensity of the bond between the input variable and neuron. In the case of multi-layer network (Fig. 2) the weight functions w_i represent the intensity of bonds between the input variable and the neuron layers. Then the output variables are predicted as weighted linear combinations of the outputs from the last internal neurons layer (hidden layer). In this case each neuron can simultaneously receive and transmit any finite number of different information signals. Therefore NN is strictly speaking non-parametric non-linear regression. Neural

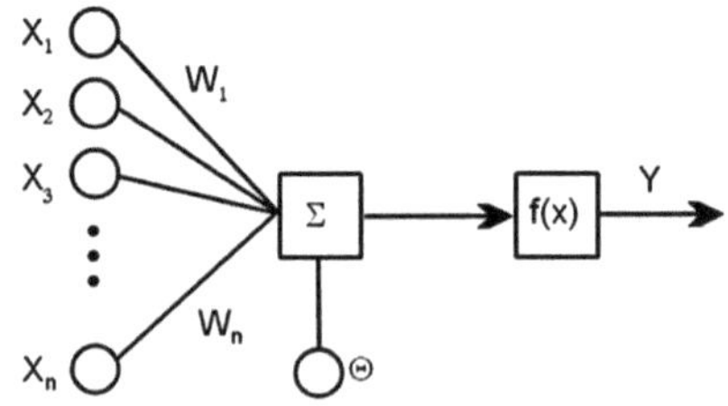

Fig. 1. Mathematical model of neuron

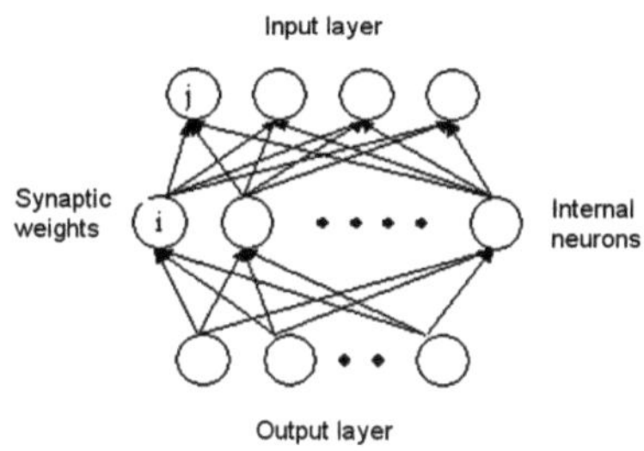

Fig. 2. Structure of multi-layer network

networks are also divided into a feed forward (non-recurrent) and a recurrent one. In the feed forward NN the information flow is univocally defined and so the identical input data (stimulus) always gives the same response of the network. Recurrent network response is not given only by actual input data (stimulus) but the time influence on the input data is also implemented (using recurrent neurons).

Artificial neural networks operate in two phases: optimizing (adaptive, learning) and predicting (active). The network learns to solve the problem in optimizing phase. NN learns mostly by using iterative process of changing the weight factors (synaptic weights), alternatively by adjusting the activation (transmission) function or by changing the network topology. NN is optimized using the least squares method, i.e. that the sum of squares of differences between prediction and the measured output variables values was minimized. We distinguish between two types of NN learning:

- A type with "a teacher" involved: a training set (a teacher) represents the desired output. Network learns by comparing its output to the desired output and by changing the synaptic weights it tries to minimize the difference between the actual and the desired output. There are two types of algorithms - learning using batches of data and learning sample by sample (both types are combined in nowadays practice).
- A type without "a teacher" (self-organization) - NN sets an optimizing criterion itself, mostly by evaluating statistical data between the input signals. The weights of connections are set so that the NN using the same or similar input data provides the same response (cluster analysis is used).

In the active phase NN processes new input data according to the learned settings (optimized algorithm) from the adaptive phase. The input layer of the NN becomes unbalanced after reading up the data. The memorized values in neurons will be transformed due to signals flow through connections between neurons until they have reached a stable state (equilibrium). Then there will be a response to the input data on the NN outputs [10].

A typical procedure of using neural network may be as follows:

1. A group of independent variables (predictors), which has an influence to dependent variables, has to be selected. Then a group of dependent variables is to be selected (group can contain only one variable), which should depend on the predictors.
2. The architecture of the NN has to be selected, i.e. the number of layers and the numbers of neurons in each layer. There is no general rule for creating the best

network architecture, however it is appropriate to use a number of neurons roughly corresponding to the number of variables. Single hidden layer networks are recommended in the case of linear or slightly non-linear relationship. Such NN are generally sufficient for most solved problems. Two-layer networks can be used for nonlinear relationships, but these are usually not very effective (there is a risk of overdetermined ambiguous and unstable models or difficulty with optimization). Masters [11] states that for discontinuous rock mass two layer net has to be used. The number of input data should be at least ten times higher than the number of neurons in the network. The networks with 2 – 20 neurons and 1 – 3 layers are usually sufficient architectures for large-scale problems [12].

3. The next step is optimization of the network parameters ("learning" neural networks). During this process, the optimizing algorithm tries to find a set of weight factors w_i, so that the predicted values are in the best accordance with the entered dependent variables. There is no uniform view on the choice of input weights. Some authors argue that the choice of the default values of weight factors affects convergence of the NN, others say that it doesn't. If optimizing starts with random values of weight factors w_i, each solution found by optimizing is quite different from the others. The correspondence of the dependent input data is measured by the least squares method. Though the settings of the weights in a net is completely different, the model can provide almost identical prediction with the equal sum of squares for the given input values. It cannot be generally guaranteed that the solution is the best possible. Therefore, it is advisable to run optimization several times to get a better residual sum of squares. The difference in optimization of NN can, however, sometimes influence the quality of predictions for unknown data.
4. The successfulness of neural networks can be assessed by validation (cross validation). In this technique we select 70% of the input data (this value is not mandatory) for the validation. The remaining 30% of input data are then used for optimization. The prediction (calculated after or during optimization) is compared with the actual value response. If this prediction is close to the actual response, it is confirmed that the NN has ability to predict the response for unknown data correctly. Quality of prediction can be assessed using a plot of the training (network optimization) process or prediction graph (Fig. 3).
5. The last step is to use the optimized network for prediction of unknown values. The new values of the independent network input variables are set as in article 1.

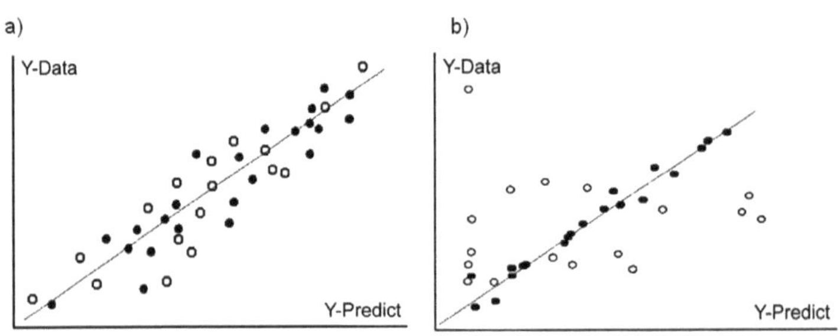

Fig. 3. Data-prediction plot (QCExpert) (a) Good prediction capability (b) Poor prediction capability of the network

3 Back Analysis of the Tunnel

The Brusnice tunnel is a part of the Blanka tunnel complex in the north-western section of the city ring road in Prague. The tunnel was driven using New Austrian Tunnelling Method (NATM) from 2009 to 2011 in the geological environment of the Prague Basin, where a complex of aleuropelitic shale, greywacke, sandstones and quartzites of the Early Palaeozoic age form the bedrock. The design of final lining consisted of reinforced concrete with membrane insulation against pressure water. Minimum thickness of the invert was 600 mm (concrete of C30/37-XC1 grade); the thickness of the crown was at least 550 mm (concrete C30/37-XF2 grade with polypropylene fibers 1 kg per 1 m^3). The concrete lining was reinforced with steel bars and welded mesh KARI class B500 B. Selected cross section for back analysis using NN is situated 220 m from the Myslbekova portal where the excavation started. The tunnel consists of northern and southern tunnel tube (16.5 m wide and 12.7 m high) with excavated cross section of about 169 m^2. The overburden is approximately from 25 m to 26 m thick.

A mathematical model of the tunnel and its vicinity was made in software PLAXIS 2D V8 (Fig. 4). Mohr-Coulomb constitutive model was adopted to describe the behaviour of ground. Tunnel lining was modelled by one-dimensional beams placed in the axis of the primary lining (shotcrete 35 cm, 15GPa.). 3D behaviour of the rock mass was introduced into the model using β-method.

The first step in creating the NN architecture is the input and output data determination. The ranges of material parameters of ground situated in the selected cross section are considered as input data: unit weight γ (24, 5 – 27, 0 [kN/m^3]), cohesion c (100 – 250 [kPa]), angle of internal friction φ (30 – 41 [°]), Poisson's ratio ν (0, 20 – 0, 30), elastic modulus E (300 – 700 [MPa]), lateral pressure coefficient K_o.

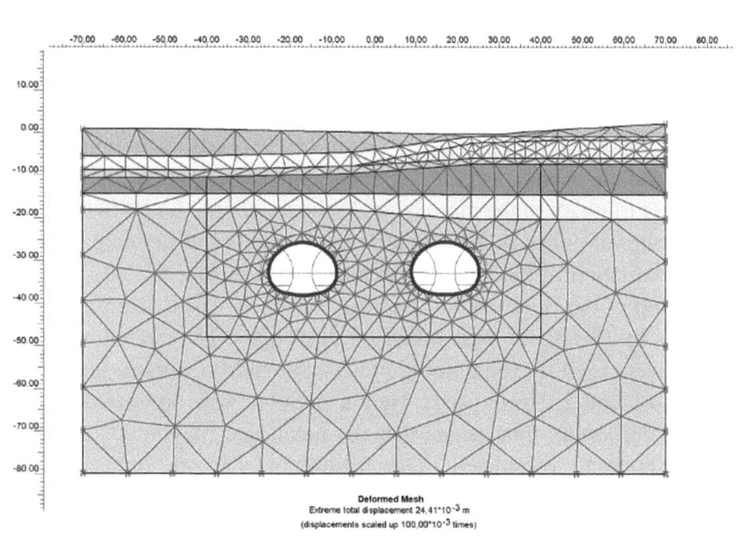

Fig. 4. Numerical model (geometry and finite element mesh)

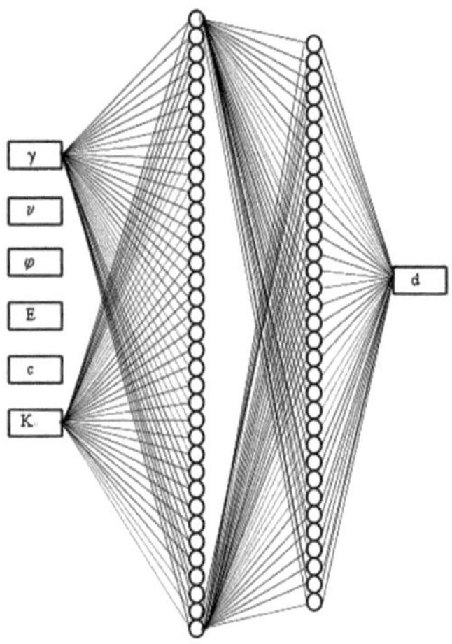

Fig. 5. Architecture of ANN

The output parameters are displacement in the crown and maximal surface settlement. The network topology consists of four layers (input, output and two hidden ones). The input layer transmits the input data signal to all neurons in the next layer (hidden). It ensures that there is in each layer a complete connection. There are recommendations regarding the number of neurons in the network architecture, e.g. the number of neurons in the hidden layers should be at least two times higher than the number of neurons in the input layer. However, the number of elements in the network is estimated in practice and then it is adapted according to the results (it is recommended to make an experiment with the number of perceptrons in the hidden layers). It is also necessary to determine the number of steps in the iterative optimization process. The parameters moment and learning speed affect the optimization algorithm too (recommended values are from 0.1 to 0.9). Further, the activation function (continuous and differentiable) has to be determined. In our case a sigmoid with steepness equal to 1 was selected as activation function. This parameter indicates the sensitivity (speed) with which the neurons respond to the change of independent variables. The parameters of the NN architecture were: the number of nodes in the input layer: 9; the number of nodes in the hidden layers 36 and 30; number of nodes in the output layer 1; number of iteration steps 3000. The graphical representation of the network architecture is in Fig. 5.

The next step is the optimization process, which is called learning or training of neural networks by finding the best values of all weights. The random values of weight coefficients w_i were used for the training process (a random number function (RND) in the range of −0.1 and 1.0 was used). Neural network was optimized to satisfy least residual squares criterion, this means that the mean and maximum error for learning and testing data were calculated. The results of network optimization are: the structure 9-36-30-1; learning speed 0.3; moment 0.9; terminating when error < 0.05; maximum error for learning data 0.0026; maximum error for verification data 0.0042. The graph of an optimization process was used to assess the quality of the ANN

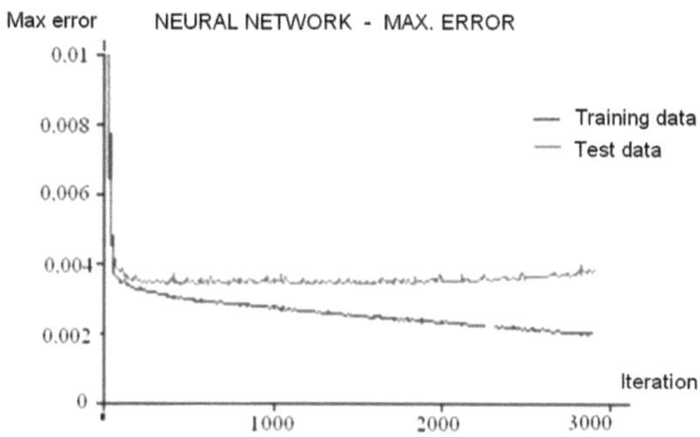

Fig. 6. ANN optimization process

Table 1. Comparison of measured and calculated values

	In situ measuremets	FEM model	Prediction NN
Max. surface settlement	33 [mm]	26 [mm]	28 [mm]
Tunnel crown displacement	49 [mm]	38 [mm]	42 [mm]

prediction (number of iterations versus sum of squares of the differences between predictions and the measured output variables) – Fig. 6.

Two measurement points were chosen for the back analysis: the vertical deformation on the surface and displacement in the crown of both tunnel tubes. Table 1 shows the comparison of measured, computed (FEM) and predicted values (NN).

4 Conclusion

There are many tunnel designs performed by numerical analysis based on the finite element method in geotechnical engineering practice. It is generally known that the soil and rock parameter obtained as a result of a geotechnical survey are susceptible to errors that it may affect the calculated values. To ensure the best agreement of the numerical model behaviour and the reality the back analysis can be used. This procedure is relatively simple (therefore engineering prefer it), however, it does not necessarily lead to the desired results. Between the other back analysis methods the artificial neural network method is the one which is used the most in the highly developed tunnelling countries. It has the following advantages:

- It can take into account the accuracy of the measured data by determination the weight factors of the neural network.
- Commercial software includes instruments that lead the user through all the necessary steps of creating the neural networks.

Based on the obtained results it is possible to state that the neural networks are useful tool for prediction of the tunnel deformation and the ground surface settlement. A good level of accuracy was achieved after optimization of the neural network.

Acknowledgement. This article was written as part of the task of the project supported by Competence Centres program of Technology Agency of the Czech Republic (TA CR), no. TE01020168.

References

1. Cividini, A., Jurina, G., Gioda, G.: Some aspects of characterization problems in geomechanic. Int. J. Rock Mech. Min. Sci. **18**, 487–503 (1981)
2. Lee, C., Sterling, R.: Identifying probable failure modes for underground openings using a neural network. Int. J. Rock Mech. Min. Sci. Geomech. Abs. **29**, 49–67 (1992)
3. Moon, H.K., Na, S., Lee, C.W.: Artificial neural-network integrated with expert system for preliminary design of tunnels and slopes. In: Proceedings of the 8th International Congress on Rock Mechanics, Balkema, pp. 901–905 (1995)
4. Yoo, C., Kim, J.M.: Tunnelling performance prediction using an integrated GIS and neural network. Comput. Geotech. **34**, 19–30 (2001)
5. Park, H.: Study for Application of Artificial Neural Networks in Geotechnical Problems. Korea Republic, Samsung C&TI (2011)
6. Ochmański, M., Bzówka, J.: Back analysis of SCL tunnels based on artificial neural network. Archit. Civil Eng. Environ. **3**, 73–82 (2012)
7. Li, C., Feng, J.W., Zhao, K.: Back analysis of rock parameters based on neural network and FLAC. In: Yang, C. (ed.). Transit Development in Rock Mechanics. Taylor and Francis (2015)
8. Zapletal, A.: Kritické zhodnocení možností zpětné analýzy. Tunel **2**, 52–60 (2011)
9. Pruška, J.: Využití neuronových sítí pro zpětnou analýzu při návrhu tunelů. In: Tunely a podzemné stavby 2015, STA Žilina (2015)
10. Rojas, R.: Neural networks: A Systematic Introduction, 1st edn. Springer, Berlin (1996)
11. Masters, T.: Practical Neural Networks Recipes in C. Academic Press, Boston MA (1994)
12. Information. http://www.trilobyte.cz

Some Design Considerations of a Fire Within a Sub-surface Railway Switches and Crossings

Aaron McDaid[1(✉)] and Nicole Hoffmann[2]

[1] Parsons Brinckerhoff, London, UK
Aaron.Mcdaid@pbworld.com
[2] Kingfell Ltd, London, UK

Abstract. This study was performed to assess the event of a fire on a train within the S&C area, addressing the limiting of smoke flowing into either the adjacent station or the tunnels. Both areas of infrastructure had to be protected as they could be occupied by people either at the station or in trains, as well as being potentially part of the evacuation route for occupants of the S&C area. By the very nature of the S&C, the evacuation of passengers within this area also posed an issue due to the routes to safety, which are not easily protected in such a relative large enclosed space. The access routes for fire brigade and the fire-fighting environment were also considered. The following briefly outlines the application of two fire engineering toolbox techniques; CFD smoke modelling and evacuation analysis. These tools provide important components required to perform an ASET and RSET assessment, which can provide powerful insight into the ability of passengers to escape. By constructing a 3D CFD model and measuring key parameters with respect to time, the tenability along the potential escape routes was considered. Combining this data with a travel time assessment of the evacuating passengers informed the performance-based approach to develop an optimum and specific ventilation and evacuation strategy. These in turn allow for guidance to be developed on the training and management requirements, which are to form part of the overall system's operational emergency procedures. It also addressed the considerations for the access provisions for the fire brigade.

Keywords: Smoke modelling · CFD modelling · Evacuation modelling · Ventilation · Tenability · Movement in smoke

1 Introduction

There has been a lot of research and development of smoke ventilation systems and strategies for stations, tunnels and underground environments. When these are interconnected, all the skills and experience from previous studies are combined to develop an alternative design solution. One of our recent projects concerned the provision of a new rail route consisting of three bored tunnels, a sub-surface station and an area of switches and crossing (S&C). An S&C area allows trains to interchange between the three tunnels and the station platforms to provide operational flexibility. Though the S&C could be open to atmosphere, existing ground level infrastructure did not make this an option, creating a very complex interconnecting set of underground

A. Kravcov et al. (eds.), *Durability of Critical Infrastructure, Monitoring and Testing*,
Lecture Notes in Mechanical Engineering, DOI 10.1007/978-981-10-3247-9_5

environments. In terms of ventilation systems, the station was to be naturally ventilated via roof level openings, the tunnel were to be mechanically ventilated via shafts within the tunnels and the S&C area was to be provided with a mechanical smoke extract system. From a fire safety point of view, the S&C area has to be assessed just like any other section of the rail infrastructure as to the provisions required in the event of a fire incident. In the event of a fire on a train, the train follows usual procedures continuing to the next station to expedite the evacuation via the platforms. Stopping a train within a tunnel is usually considered a last resort. The development of evacuation and smoke ventilation strategies are more critical in that environment. Station occupants are to be evacuated via the evacuation routes within the station only, the tunnel occupants are to cross through into one of the adjacent bore tunnels and from there be evacuated; whilst the S&C area was to have its own protected routes direct to the outside of the sub-surface environment, as well as routes into the adjacent station and tunnels. The latter very much dependant on the location of the fire and the associated ventilation strategies.

Another important design requirement was the minimisation of the impact of a fire within the S&C area on the adjacent station. To address all the requirements and have a holistic overview of the issues, a deterministic approach was adopted. This involved developing a series of scenarios simulated using Fire Dynamics Simulator (FDS) for both the smoke and the evacuation modelling [1, 2]. These studies showed the effectiveness of a smoke exhaust system as well as the interaction with the station and tunnel ventilation systems. These provided insight into the tenability of the S&C area in terms of maximum temperature and visibility reduction, in turn incorporated within the separate FDS + Evac evacuation model. Considerations for the varying passenger travel speeds due to impact of smoke were also assessed.

2 Scenario

The S&C area is a relatively large enclosure, measuring approximately 300 m in length, 72 m in width and 15 m in height, accommodating a ventilation plant at high level along with associated ductwork (see Fig. 1). The ventilation plant location separates the area into three principle zones: left zone, underneath the plant zone and right zone.

The station is to be located on the western side with the three running tunnel bores on the eastern side. The sub-surface platforms of the station will be ramped to allow passengers access from track level to the protected escape staircase located at the end of the platforms. Along the interface between the station and the S&C area are station accommodation levels, forming part of the smoke reservoir within the S&C area. This reservoir is bounded on the tunnel portal side by the height differential between the tunnel bores and the roof level.

The trains are approximately 200 m in length, allowing potential for the complete train to be located within the S&C area, requiring passengers to detrain to track level via high level walkways. Several potential locations for a fire incident within this area have been considered within the wider study. Here, only one of the scenarios is considered, with a fire on a single carriage. The incident carriage is assumed to be located within the left zone, nearest to the station. The S&C ventilation plant located 10 m above track level is inactive at the beginning of the simulation. After 180 s the S&C ventilation plant is activated and ramped up over 30 s to 280 m^3/s after which it remains constant. This is to account for the time required for the ventilation operator to activate the correct ventilation mode. The tunnel ventilation plant was assumed to be inactive within this simulation. However, an induced air flow is used to replicate the tunnel flow conditions. The natural ventilation openings in the station are assumed to open after 180 s. For the evacuation of the incident train, it is assumed that passengers are directed to evacuate only via the last door of the end carriages. This is to maximise their time within the relative protection of the carriages, whilst minimising their time and their travel distances within the S&C area. During the simulation, monitoring points were located 2.5 m above the track and 1 m away from the train, to collate tenability values along their evacuation routes. The study evaluates the condition for fully developed design fire 1,000 s after the start of the simulation.

3 Smoke Modelling

Smoke movement was simulated using FDS MPI v. 5.5.3 [1], whereby the smoke is treated as an inert mixture of air and homogeneous smoke. Fluid turbulence modelling is performed by the application of the large eddy simulator (LES) model, which is based on the Navier-Stokes equation.

3.1 Model Geometry

The S&C area along with the section of the station box has been modelled to consider not only the area of interest but also the influence of the natural ventilation openings in the station. Only the nearest natural ventilation openings have been included in the model geometry. In Fig. 1, the four squares denote the natural ventilation openings to atmosphere within the station box. The tunnel portals are denoted at the right hand side along with the S&C mechanical ventilation plant and the ventilation ducts.

It was assumed that windows on the train fail at 300°C, allowing smoke to vent directly into the S&C area.

Computational Mesh. Due to the size of the simulation volume, a cell size of 0.5 m was adopted for the scenarios. This was the smallest cell size that could be used due to computational limitations. The characteristic fire diameter (D*) [1] was calculated using Eq. (1) as 2.08. This gives a D*/dx ratio of 4.1 which is within the bounds of 4 to 16 in accordance with NUREG 1824 [3].

$$D^* = \left(\frac{\dot{Q}}{\rho = c_p \, T = \sqrt{g}} \right)^{\frac{2}{5}} \quad (1)$$

Smoke Reservoir. The S&C area is a high space bounded by the station box and the tunnel portal construction. By using the maximum height available between the track

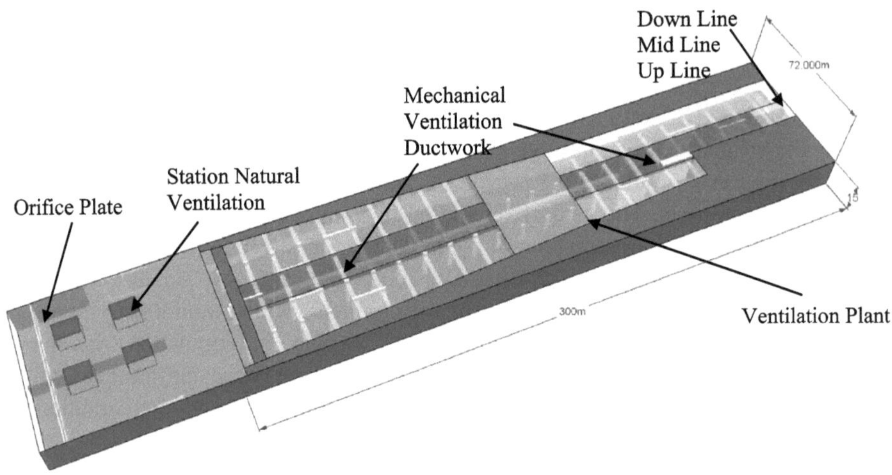

Fig. 1. Switches and crossings CFD model

and the ground level slab, a smoke reservoir is created within the S&C area. As it is approximately 80 % of the S&C area, it is to assist in containing the smoke within the S&C area, and limit any potential smoke flow into the station.

S&C Ventilation. The S&C ventilation is prescribed by a velocity flow across the cross section of the vent openings, provided along the over track exhaust system within the S&C area. The ventilation system is a directional system which allows the extract to be focused either side of the ventilation plant area.

It is assumed that the S&C ventilation system activates after 3 min into the fire scenario ramping up over a further 30 s to full extract capacity. It was assumed that the air supply temperature was 30°C.

Tunnel Ventilation. Tunnel ventilation is prescribed by a velocity flow across the cross section of each tunnel bore induced by the S&C ventilation system. The boundary conditions at the tunnel portals were obtained from the tunnel ventilation design team from the 1D smoke control analysis using Subway Environment Simulation [4]. These values were calculated at 30 m^3/s per tunnel bore. It was assumed that the air supply temperature from the tunnel was 30°C.

Station Ventilation. In order to replicate the aerodynamic resistance of the station, an orifice plate of 60 % of the vertical cross sectional area was placed on the station side. This was used as it was not possible to simulate the total station geometry. It was assumed that the air supply temperature from the station was 30°C.

Design Fire. The fire was assumed to be fully developed at the start of the simulation within the S&C area, as it was assumed that the fire grew during the travel period within the tunnel. The peak heat release rate was assumed to be 7 MW.

The total heat release rate was distributed over the portion of the floor area of the incident carriage measuring approximately 8 m^2. This resulted in a heat release rate per unit area of 875 kW/m^2. The dimensionless heat release rate $(\dot{Q}*)$ [5] was calculated using Eq. (2) as 0.34, which is within the reasonable bounds of 0.1 to 2.5 [6].

$$\dot{Q}^* = \frac{\dot{Q}}{\rho_\infty c_p T_\infty \sqrt{gD} D^2} \tag{2}$$

The fuel was selected as polystyrene, which is widely considered to be a very dirty fuel. This fuel was modeled with a soot yield of 0.164 g/g [7] with a heat of combustion of 27 MJ/kg.

4 Evacuation Modelling

The evacuation was simulated using FDS + EVAC v. 2.3.1 [2]. The 200 m long train has a crush loaded capacity of 550 people distributed over 11 carriages. It was assumed that it would require 180 s to initiate the evacuation. Due to the nature of the evacuation it was also assumed that additional time would be required for passengers to commence movement which was naturally distributed over the population from 180 s to 240 s. It was assumed that the occupants would transfer from the train or walkways to track level via high level walkways which were strategically located throughout the S&C area to limit the stepping distance for the varying abilities of the passengers.

Two exit doors were assumed to be in use for the evacuation off the train, as members of trained staff would be present to assist in the evacuation. Two pressurised evacuation shafts were located at each end of the S&C area; see Figs. 2 and 3.

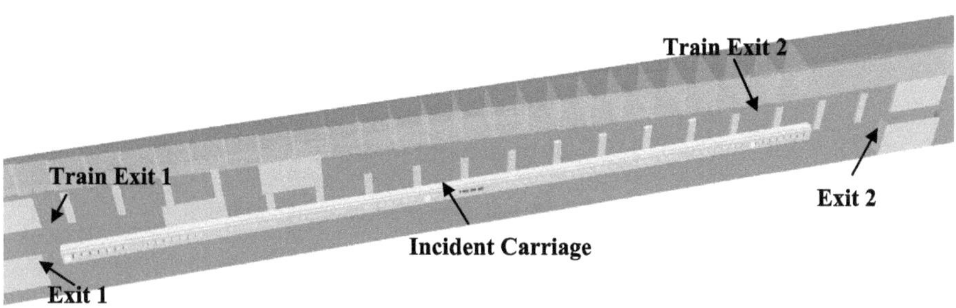

Fig. 2. Scenario layout

Effect of Smoke on Passengers. There is limited body of research with a wide variation in results related to the movement of people in smoke. Research by Jin [8] suggests that there is a reduction in speed at very low smoke densities. More recent research by Frantzich and Nilsson [9] suggests that the speed reduces at a much slower rate; see Fig. 4.

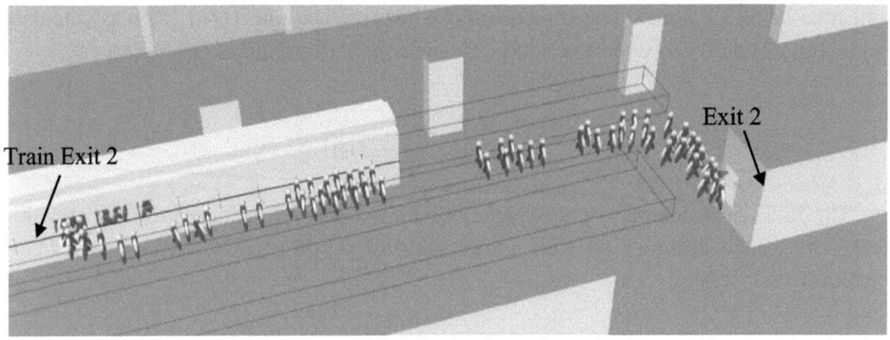

Fig. 3. Example of evacuation

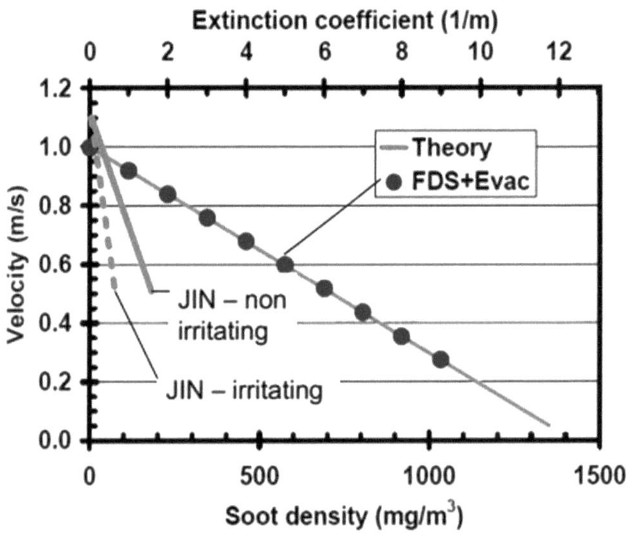

Fig. 4. Comparison of walking speed in smoke research

The design walking speed was naturally distributed between bounds of 0.5–1.2 m/s [10]. The evacuation minimum walking speed was capped at 0.3 m/s as shown in research by Jin [8].

This difference in walking speed reduction may be due to the difference in irritant concentration of the gasses between the two data sets; it may be due to experimental differences; it may be due to interpretation of the data. The range of scatter is noteworthy in the data for the Frantzich and Nilsson research [9]. Figure 3 shows the data side by side and at the same graph scales. The effect of smoke movement on the walking speed of agents in FDS + EVAC is controlled by data from experiments by Frantzich and Nilsson. The walking speed within FDS + EVAC uses a numerical relationship derived from these experiments as shown in Eq. (3) [9]. The walking speed is influenced by the smoke extinction coefficient which is determined by Eq. (4).

$$V_i^0(K_s) = MAX\left\{V_{i,\min}^0, \frac{V_i^0}{\alpha}(\alpha + \beta K_s)\right\} \tag{3}$$

$$K = K_m \rho Y_s \tag{4}$$

After an initial assessment of the total evacuation time it was considered that due to the irritant nature of the smoke within the enclosed environment, the reduction in walking speed may not be representative in the default data adopted in FDS. It was considered that the walking speed may reduce at a greater rate, justified by the significant difference between the Frantzich and Nilsson and the Jin data, and also Jin's observation of the significant effect of irritants. A decision was made to attempt to increase the rate at which agents slow down to account for the large amount of uncertainty in the Frantzich and Nilsson data and the potential impacts of the irritants. An assumption was made that all smoke was of an irritant nature, having a significant effect on the movement of occupants.

The approach to slow down the occupants due to smoke exposure was by recalibrating the mass extinction coefficient (K_m). This was based on the difference in slope between irritant and non-irritant smoke from the Jin data; see Fig. 5. This recalibrated value would slow the occupants down to take into consideration potentially more irritant smoke.

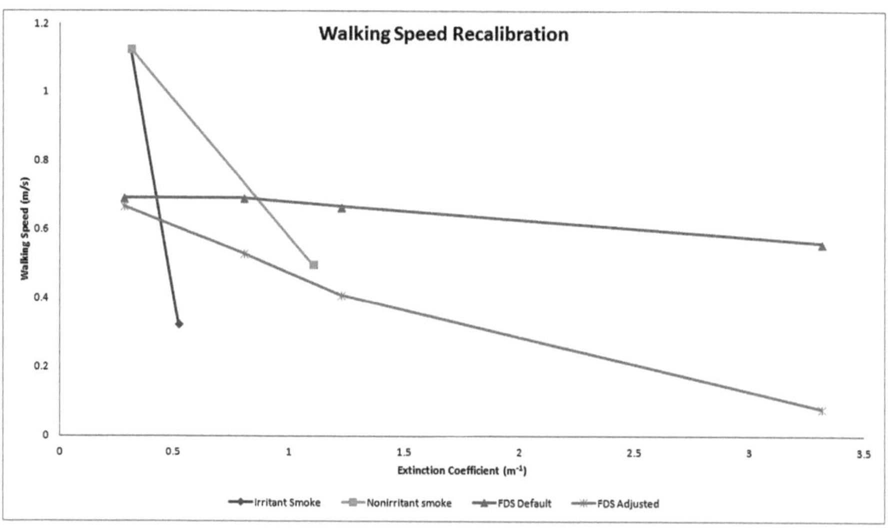

Fig. 5. Walking speed recalibration

The Jin data suggests that the walking speed decays at a rate about five times greater in irritating smoke than non-irritating smoke. It was therefore decided to slow down the FDS evacuation in smoke by a similar factor. This method was validated using a simple evacuation model and tracking walking speed on a horizontal surface for

Table 1. Walking speed recalibration calculation

Mass fraction of smoke assume	Y_s=	50 mg/m^3	0.00004172 kg/kg
Smoke density assume	ρ=	1.2 kg/m^3	1.2 kg/m^3
Calculated		8700 kg/m^2	41958 kg/m^2
Light reflecting sign	C=	3	3
Extinction coefficient	K_S	0.436 m^{-1}	2.101 m^{-1}
Visibility distance	S	6.9 m	1.4 m

a given mass fraction of smoke. This process was repeated for different smoke extinction coefficients (m^{-1}) thus giving different walking speeds. The result of the reduction is shown in Table 1.

It was found that by adopting a specific extinction coefficient of 41,958 kg/m^2, rather than the 8,700 kg/m^2 default in FDS, the walking speed could be reduced significantly (by a factor of approximately five). A consequence of this is that the visibility data presented in the FDS outputs is no longer the 'real' visibility. It has been termed 'effective visibility' and is about five times lower than would be expected for only the visibility effect. For example, if the FDS outputs show 2 m effective visibility, the 'real' visibility would be expected to be 10 m. To overcome this, the actual visibility was recalculated manually my placing devices measuring the key inputs required.

Performance Criteria. The performances of the smoke ventilation systems were assessed on the tenability of the S&C area and station, which will form part of the evacuation routes.

The tenability of the evacuation routes were based on the overall conditions which were monitored 2.5 m above the track. Irritants and toxins produced by the fire are likely to have significant effect on the conditions in the S&C area. However, as currently detailed data is not available in relation to the trains to be used, any assumed distribution of irritants and toxins would be subject to a significant margin of error. On this basis, consideration of this issue is left as further work. Additionally, tenable conditions should be maintained for as long as feasible, to assist attendance by the fire brigade.

Thermal Tenability. Thermal tenability is described as the limit exposure concentration for asphyxiant gasses for 1,000 s. This is based on a minimum clear layer of 2.5 m above the tack level.

PD 7974 Part 6 [11] also states that convective heat for a smoke layer above 60°C will make an escape route impassable. Temperature data is shown with bounds of 30°C (ambient) and 60°C (convective tenability limit).

Smoke Layer Temperature. In addition to thermal tenability, is the radiant heat exposure of people from a hot smoke layer above the clear layer height. PD 7974 Part 6 [11] states that the threshold value for radiant heat gain from a smoke layer is 200°C.

Visual Tenability. The criterion for tenability is taken from PD 7974 Part 6 [11], where visibility in smoke less than 10m will result in evacuees regarding the route as unsafe. Visibility and obscuration is calculated in FDS using Bougers Law as per Sect. 2 of the SFPE Handbook [10]. Visibility data is shown with bounds of 0 m (no

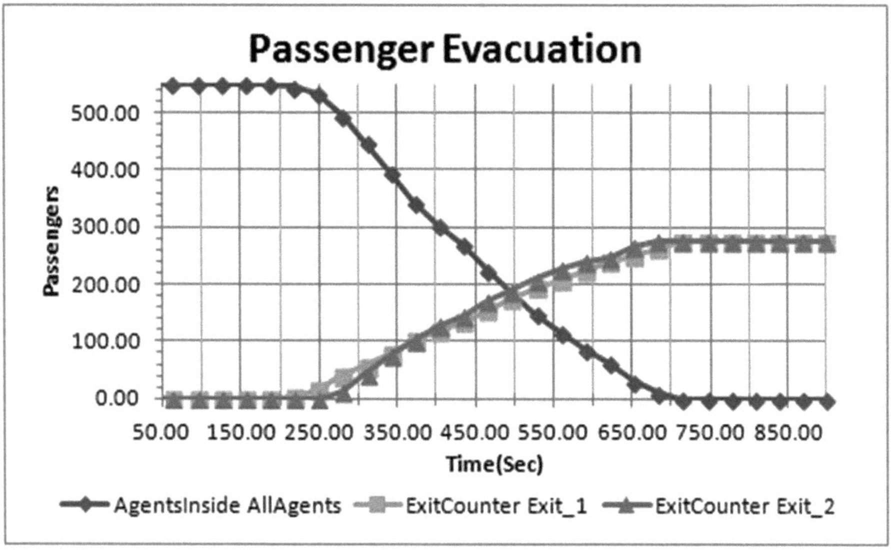

Fig. 6. Passenger evacuation timing

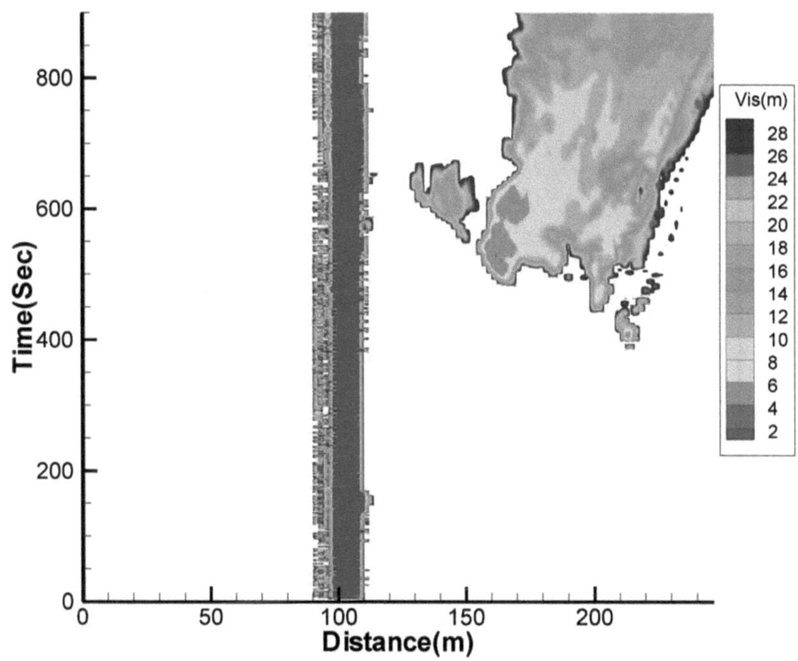

Fig. 7. Effective visibility contour map at 2.5 m above track

visibility) and 30 m (clear air). The visibility contours will be presented at a height 2.5 m above track level. This is to assist to gain an insight into the movement of smoke.

5 Results

As can be seen in Fig. 6, passengers would require 710 s to evacuate the train and S&C area. The reduction in walking speed due to the recalibration of the mass extinction coefficient resulted in increasing the total evacuation time.

During that period, the smoke would be starting to fill the S&C area. The contour plot of the effective visibility (see Fig. 7) indicates that the smoke would be as low as the 2.5 m monitoring location after about 400 s. However, this is the effective visibility which is significantly worse than the 'real' visibility, which is shown in Fig. 8. This clearly indicates that there are only a few minor periods in which the smoke would have an impact on the visibility. However, even that is not significant to affect the evacuation of the passengers.

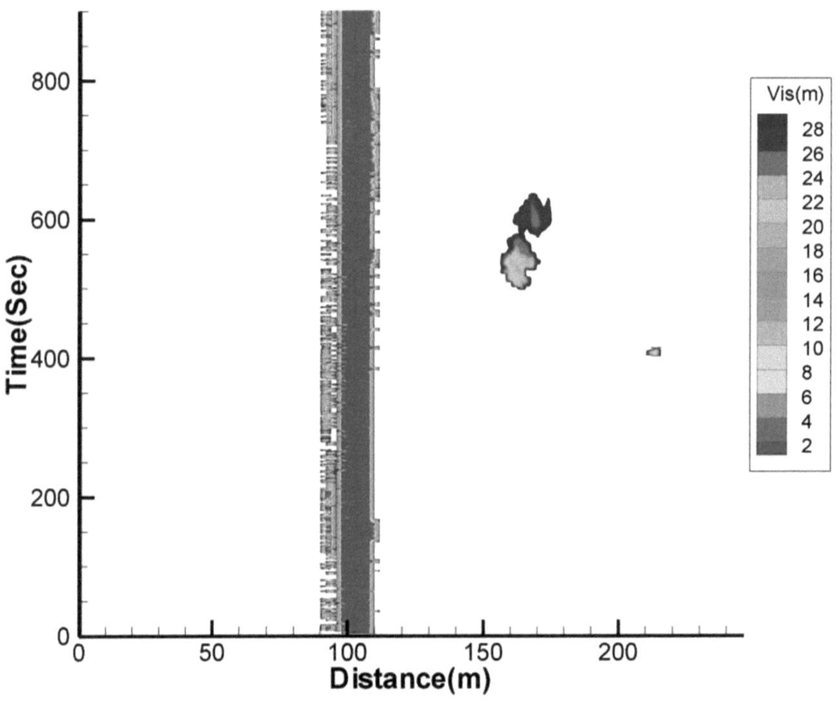

Fig. 8. Real visibility contour map at 2.5 m above track

Similar results and conclusions can be drawn from Fig. 9 which is monitoring the temperature along the evacuation route. This shows that the temperatures are not significantly higher than the tenability limit.

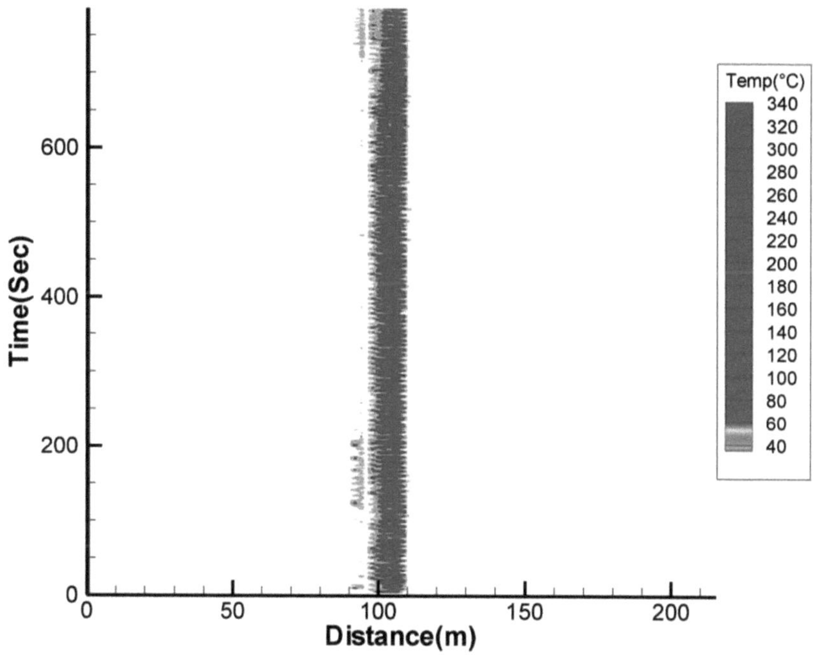

Fig. 9. Temperature contour map at 2.5 m above track

The overall results indicate that thermal and visual tenability was satisfied for all escape routes. Similarly, the simulation indicated that the smoke layer temperature would be insufficient to cause a radiant heat hazard to passengers, staff or emergency services personnel anywhere on the S&C area. Radiation tenability was only critical very near to incident carriage which would be expected.

6 Conclusion

The connectivity of tunnels, a station and the S&C area is a non-trivial design configuration from a fire engineering point of view. The above presented only a brief snapshot of some of the analysis that had been performed. These ensure that the design has taken into account the smoke production and extraction rates, staff operational times of systems and evacuation assistance, the required evacuation time for passengers to reach a place of relative safety and limiting smoke ingress into the station.

The analysis also provided further insight into the training and management requirements to achieve the safe evacuation. It was also noted the criticality of the early activation of the S&C ventilation system.

As indicated, further work is required as the project progresses such as the provision of information regarding rolling stock materials involved in a fire.

References

1. Federal Emergency Management Agency FEMA: FEMA 154: Rapid Visual Screening of Buildings for Potential Seismic Hazards: A Handbook. Washington DC, USA (2002)
2. GNDT 2007: Manuale per il rilevamento della vulnerabilità sismica degli edifici - Istruzioni per la compilazione della scheda di 2° livello (2007)
3. Mcgrattan, K.B.: Fire Dynamics Simulator (Version 5) User Guide. NIST Special Publication pp. 35,163, March 2006
4. Korhonen, T., Hostikka, S.: Fire Dynamics Simulator with Evacuation (FDS + Evac), Technical Reference and User's Guide, pp. 17–30 (2010)
5. Hill, K., Dreisbach, J., Joglar, F., Najafi, B., Mcgrattan, K., Peacock, R., Hamins, A.: Verification and Validation of Selected Fire Models for Nuclear Power Plant Applications, p. 35. NUREG 1824, United States Nuclear Regulatory Commission, Washington, DC (2007)
6. Elpidoru, D.P, Huang, G.D., Karra, M.J., Kennedy, W.D., Lambrou, A., Lemmon, G., O'dwyer, T.P., Huang, G.G., Li, S.K.L.: Subway Environmental Simulation 2000 Users Manual, p. 46. Parsons Brinckerhoff, New York (2003)
7. Karlsson, B., Quintiere, J.: Enclosure Fire Dynamics. CRC Press, Boca Raton, pp. 46–79, 239 (1999)
8. Cox, G.: Combustion Fundamentals of Fire. Academic Press, London (1995)
9. Liptrott, S., Marshall, S., Hopkinson, J.: CIBSE Guide E, p. 51, 3rd ed. The Charlesworth Group, London (2010)
10. Jin, T.: Visibility through fire smoke. J. Fire Flammability **9**, 135–155 (1978)
11. Frantzich, H., Nilsson, D.: Utrymning genom tät rök: beteende och förflyttning, 75 p., Report 3126. Department of Fire Safety Engineering, Lund University, Sweden (2003)
12. Dinenno, P.J., Drysdale, D., Beyler, C.L., Walton, W.D., Custer, R.L.P., Hall, J.R., Watts, J.M.: SFPE Handbook of Fire Protection Engineering, 3rd edn, pp. 247–258. National Fire Protection Association, Massachusetts (2002)
13. British Standards Institute, PD 7974 Part 6: Human factors: Life Safety Strategies – Occupant Evacuation, Behavior and Condition (Sub-system 6), British Standards Institute, p. 37 (2004)

Replacement of Permanent Bridges After Floods in 2013

Martin Benda(✉)

University of Defence Brno, FVT, Kounicova 65, 66210 Brno, Czech Republic
martin.benda@unob.cz

Abstract. The paper deals about operation of Czech Army Engineer Corps in year 2013. There were three units of Czech Army participate in this operation. University of Defence, Military Geodetical and Hydrometeorological Office and 15 Engineer Brigade.

The aim of this operation was to construct temporary bridge after flood. During this operation was used REACH-BACK concept between University of Defence and Geodetical and Hydrometeorological Office and between University of Defence and 15 Engineer Brigade. Communication and cooperation between these units was supported by Information portal of engineer corps. The portal works as a support element for REACH-BACK concept.

Keywords: Reach-Back · Engineer · Temporary bridges · TMS and MS bridge type

1 Introduction

The Czech Republic territory was hit by floods in June 2013. There were damaged bridges mainly in middle part of part of Czech near the Prague. The engineer experts from University of Defence have essential experience with overcoming the obstacles caused by flood. They designed 20 bridges in the disaster affected areas of the Czech Republic in 2010 [6, 10]. The most of them were bridges from MS set but one the longest and the most difficult was bridge from TMS set (Fig. 1).

Last year the soldiers from battalions 15 Engineer Brigade erected 4 MS type bridges and soldiers from Bridge Company 151 Engineer Battalion built the bridge TMS. The operation of designing and constructing of temporary bridge TMS took almost 5 weeks (from end of June to start of August).

There were three units of Czech Army participate in this operation - University of Defence (Department of Engineer Technology), Military Geodetical and Hydrometeorological Office, and 15 Engineer Brigade. Every mentioned unit organized particular team or teams that were included in this operation.

REACH-BACK concept was used for the first time as support for so large operation in 2009. Basic principles of the concept demonstrated that it can operate successfully in praxis. That was the reason why the concept REACH-BACK together with Information Portal of Engineer Corps (IPEC) was used for support during this operation again.

It is suitable to introduce the Reach-Back concept at first and explain the main possibility of its usage. It is a new model of cooperation and communication between

A. Kravcov et al. (eds.), *Durability of Critical Infrastructure, Monitoring and Testing*,
Lecture Notes in Mechanical Engineering, DOI 10.1007/978-981-10-3247-9_6

Fig. 1. Destroyed bridge construction in the village Červený Hrádek

the units deployed directly in action and main corps supporting them. This concept supposes units used for tasks compliance (for example in foreign mission) and these units do not necessary consist of all components as well as home. The components staying "home" fill tasks for "in theatre" units as their support team [1].

2 Temporary Bridging Sets MS and TMS

As a suitable temporary bridge for overbridging four gaps was chosen MS bridge set [9]. Only one bridge was replaced by TMS set because of higher carrying capacity requirement. There were damaged bridges of short span mainly. That is why most of them were replaced by temporary bridge MS type. The longest MS type bridge in Nový Knín village had span 24 m, but for launching was used longer construction (27 m). The requirements of civilian authority for carrying capacity were the next reason. The most of them needed carrying capacity up to 20 ton. Temporary bridge from TMS set was built up in Chlum u Sedlčan village and was only 15 m long.

Bridging set MS ant TMS are standardized portable steel bridges with two primary truss and lower bridge deck. It is used only for one-way traffic lane with maximum carrying capacity 60 ton. It is the most suitable to build one span bridge with length 21 m (carrying capacity 60 t) for MS type and one span bridge with length 36 m (carrying capacity 70 t) for TMS type [2, 3]. It is possible to build a bridge with longer

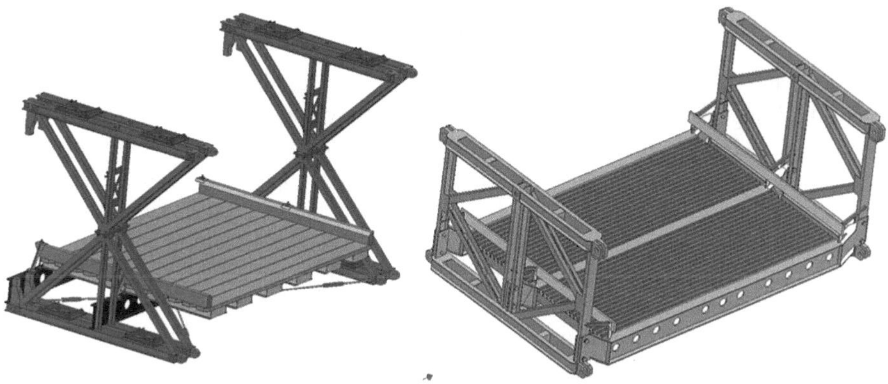

Fig. 2. Model of bay TMS and MS bridge set created by Autodesk Inventor SW

Fig. 3. Assembling of the bridges

span but the carrying capacity must be reduced. When using pier (e.g. PIŽMO), it is possible to build also bridges with more spans. The 3 m long bay is the basic assembly element for both sets (Fig. 2). It is possible do assemblage for both sets by the help of plain rollers, rocking rollers, and cantilever. The other alternative is to assemble the bridge on flat ground and put it over the obstacle by a suitable crane. The main advantage of MS type is that it is not necessary to do special modification of banks for placing the bridge and ramp. The ramp is created by folding ramps which belong to the end bridge bay. TMS type need complex ground shaping not only for placing but also for construction. The ramp for TMS must be done by special ramp or abutment wall (Fig. 3).

Both these bridge sets are not equipment of Czech Army. These bridge sets are in storage of Ministry of Transport. The soldiers of Engineer Corps are trained in special courses in Training centre of Ministry of Transport, where they get the experience with this bridge construction and also with other bridge constructions.

3 Reconnaissance and Geodetic Surveying

The main task for reconnaissance team (reco-team) was the reconnaissance of area where the bridge was planned to be built. This team consists of experts from Military Geodetical and Hydrometeorological Office from Dobruška and one member from Department of Engineer Technology who has experience with TMS and MS sets. This team configuration is possible to use for future similar tasks. Geodesists were responsible for geodetical survey of the construction site of temporary bridge. The reco-team carried out inspection of the construction site and made photo documentation. They found out span of the obstacle and suggested which type of bridge is suitable for particular gap. All information was included in Reconnaissance Report. When the reco-team accomplished the report, they placed it on IPEC.

The geodesists were lead by commander of the reco-team and under his control they made detailed geodetical survey. The data gained from survey were essential part of the Reconnaissance Report and were attached to the report as soon as geodesists prepared them. The data consisted of coordinates of important points and construction site drawing in MicroStation software. The data were used for preparation of every project of temporary bridge.

4 Project of Temporary Bridge

The project team was put together first of all. The five-member project team was from Department of Engineer Technology at University of Defence. Two members of the team worked with geodesists as members of the reco-team and other two members designed projects. Last member was leader of the team who was responsible for communication with civilian authorities. The project team used experience from regularly practised actions in collaboration with Dobruška Military Geographic and Hydro-Meteorological Office. They improved their skills and cooperation that they gathered in 2009 and 2010 during similar operation [4]. The project team repeatedly proved that the involvement of university Reach-Back laboratory in "life operation" brings good results, making the collaboration with the geodesists from Dobruška and 15th Engineer Brigade personnel smooth and easier.

Project team took data (photos, video and geodetic surveying) and Reconnaissance Report that was on IPEC server and started work on project. They also implement suggestion of reco – team that would be suitable to use TMS bridge for this situation. The project team had to accept the requirements of civilian authorities. They had two particular requirements.

The first one was that the construction of the bridge had to be above 50 years flood water level. That is why the lower chord of the truss was placed approximately 1, 2 m above terrain. The second was that construction had to be easily attached to existing road on the both sites of the banks.

The leader of project team specified type of construction, site for construction and span of the bridge. Team created 3D model of bridge construction (Fig. 4). This 3D

Fig. 4. Model of bridge placed in photo

model was also used for making project documentation. Designing necessary ground shaping in area of construction was other part of project. Microstation situation data were converted into proper CAD format. Then team created into terrain ground shaping and other modification. Then the 3D model of bridge was placed into model of terrain.

Sometime it was very suitable placed model of bridge into photo. It made placing bridge construction in real situation more illustrated.

The last part of designing was creating the final project and drawing documentation. The final project with drawing was placed on server IPEC. Then Request to prepare bridge and assembly material from store houses of Ministry of Transport was created. The Request to prepare was sent to responsible person to Ministry of Transport by the help IPEC. They prepared entire material necessary for bridge construction in the nearest store.

5 Construction of the Bridges

As was mentioned above there were erected four bridges from MS set but only one from TMS set. Since TMS bridge construction is more complicated then MS type, commander of 151 Engineer Battalion was tasked to organize the construction of the bridge. Soldiers from this unit have the best experience with this type of construction and some of their commander's work as instructors in Ministry of Transport´s courses for construction TMS bridge in Kojetín training center. The commander of building unit read up the project and made eventually consultation with the project team with the help of web conference on IPEC (Table 1).

At the first of all it was necessary to prepare building site. The commander of each building unit made meeting with local civilian authority and explain them his

Table 1. Time table of construction process

Date	Location and type of the bridge
25.7.2013	Červený Hrádek MS – 24 m
26.7.2013	Chlum u Sedlčan TMS – 15 m
10.–11.7.2013	Nový Knín MS – 24 m
27.7.2013	Velké Číčovice MS – 15 m
30.7.2013	Zadní Třebáň MS – 18 m

Fig. 5. Storage temporary bridges

requirements (e.g. cut off power line, get down trees or remove fence). The requirements came from project and other commanders drafts (Fig. 6). When whole the building site was prepared the commander sent unit to pick up the material of bridge set from the store house and to transport it into building site (Fig. 5) [7].

The commander started building of the bridge after the unit with bridge construction came into building site. The first step was to construct the rocking rollers and plain rollers trays. The rocking rollers were placed on concrete panels. The grillage from the timber was often used. The short ramp and concrete panel were used as foundation for the grillage (Fig. 3). When the plain rollers trays were assembled the launching nose were erected on rollers. When the assemblage of the nose was complete, assemblage the first bay of the bridge followed. Then the next bays of the bridge were assembled by the same way except the end bay. When the complete construction of the bridge was launched the jack down procedure started. The most of units were equipped with hydraulic jacks with load capacity 50 tons. Unfortunately, neither of them was able to work. Jacks were from the storage of Ministry of Transport and we identified that their future using is impossible due to the life span of them. The jacks had to be replaced by the jack from vehicle accessories. These jacks had load capacity only 20 tons but it was sufficient in this case.

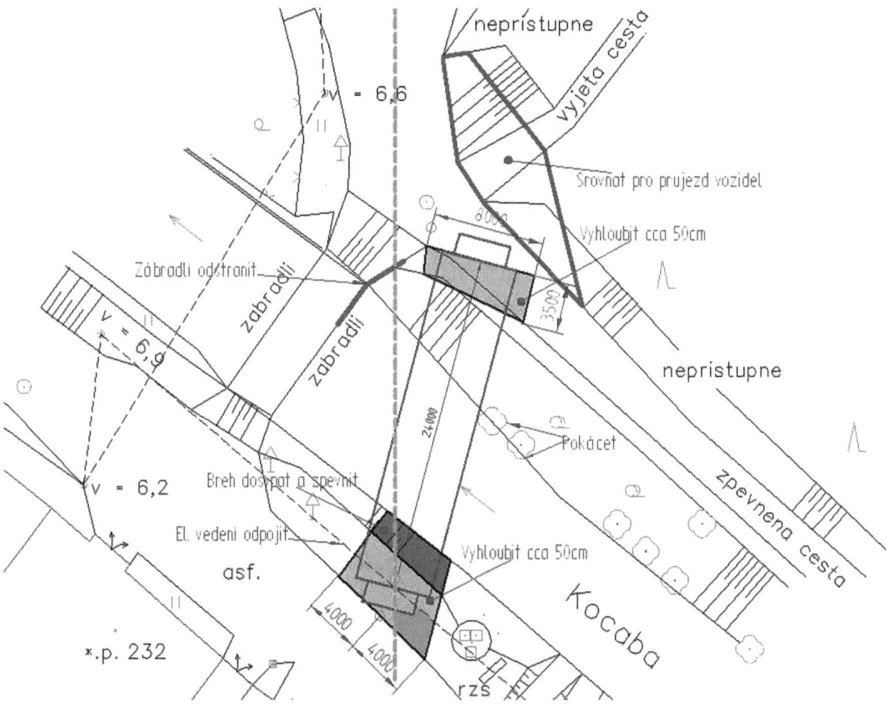

Fig. 6. Requirements for preparing the building site

Load test by using car crane was the last task and then the commander made inspection of the bridge construction [8]. The whole building process of one bridge (from picking it up from the store house to giving into traffic) took one or two day. There is not included the time that was necessary for ground shaping. The building process of TMS bridge took approximately four days because of the type and span of the bridge.

6 Summary

The range of transport infrastructure recovery by using temporary bridge was less extensive than in 2009 and 2010. All units involved in the operation "building temporary bridges" learned from the mistakes committed in last years and have made considerable progress. This progress was reached mainly in the area of MS type construction. The biggest delay in process of building was often caused by the local authorities. They often hesitated to which variants bridging decide. There were sometimes misunderstanding and it would be useful if representatives of local authorities have a general overview about temporary bridges and for what is possible and appropriate use them.

On the other hand, we have found that there are many problems that we have to cope with in the future. Some of them persisted from past. For example problems in communication between units which participate in construction of the bridge and Ministry of Transport. It was caused by the fact, that in our army there still exist units that have very rare access to internet. The problem that persisted from the past is that units are not equipped by timber that is in many cases very useful. Many soldiers participated in "live" action in 2009 and 2010 and they got valuable experience, in spite of the fact that they did not take part specialized course. These courses are organized at the training center of the Ministry of Transport in Kojetin. The most officers of construct units completed a sufficient number of courses and they are able to lead by herself building temporary bridges MS and TMS.

A significant progress was achieved in area of designing temporary constructions due to modern software tools [4]. The communication and data transmission between University of Defence and Military Geodetical and Hydrometeorological Office is on high level quality. It is due to periodical practise between these two units. It will be necessary also to determine the command and control responsibilities and to divide them between civilian and military authorities in the future [5].

Acknowledgment. The work presented in this article has been supported by the Czech Republic Ministry of Defence - K 201 Department development program "Development of technologies in weapons construction, ammunition, weapons instrumentation, materials engineering and military infrastructure".

References

1. Maňas, P., Mazal, J.: The Reach-Back concept in the czech army corps of engineers. In: International Conference on Military Technologies. Brno: Faculty of Military Technology, University of Defence (2009). ISBN 978-80-7231-648-9
2. Maňas, P., Soušek, R., Benda, M.A.kol.: Stavba provizorních mostů ze soupravy MS, 120 s. Institut Jana Pernera, o.p.s. Brno (2010). ISBN 978-80-86530-73-4
3. Maňas, P., Soušek, R. BENDA, M.A.kol.: Stavba provizorních mostů ze soupravy TMS. Institut Jana Pernera, o.p.s., 120 s. Brno (2010). ISBN 978-80-86530-74-1
4. Sobotka, J.: Comparison of elevation data of the Czech Republic for designing military constructions. Adv. Mil. Technol. **7**(2), 57–63 (2013). ISSN 1802-2308
5. Maňas, P., Soušek, R.: On Cooperation between military and civilian authorities in the Czech Republic during crisis situation in transport. In: International Conference on Engineering and Meta-Engineering icEME 2010, pp. 12–14. IIIS, International Institute of Informatics and Systemics, Orlando, Florida, USA (2010). ISBN 978-1-934272-83-1
6. Benda, M., Sobotka, J.: Construction temporary bridges after flood in 2010 – oportunity for application of Reach-Back Concept. In: ICMT 2011 - Proceedings of the International Conference on Military Technologies. Fakulta vojenských technologií, Univerzita obrany v Brně, Brno (2011). ISBN 978-80-7231-787-5
7. Použití náhradních silničních přemostění za krizových stavů. In: Novák, M. (ed.) (2009). www.sshr.cz [online]. vyd. [cit. 2012-06-06]. Dostupné z: www.sshr.cz/cinnosti/dokumenty…/Nahradni_silnicni_premosteni.ppt
8. STANAG 2021 Military load classification of bridges, ferries and vehicles. NATO/PFP (2010)

9. TP 90 Používání provizorních mostů z mostové soupravy z MS v civilním sektoru +Dodatek 1. MDS ČR, odbor pozemních komunikací, vypracoval Pontex Praha, 1996, 2010. Schváleno MDS-OPK č.j. 24911/96-120 ze dne 27.12.1996
10. BENDA, M.: Application of Reach-Back concept for construction temporary bridges after flood in 2009. In: ICMT 2010 – IDEB 2010, Bratislava, Incheba Expo, 4 s (2010). ISBN: 978-80-8075-454-9

On Utility Assessment of Building Projects

Čeněk Jarský(✉)

Department of Technology of Structures, Faculty of Civil Engineering,
Czech Technical University in Prague, Prague, Czech Republic
jarsky@contec.cz

Abstract. The paper describes the methodology of computer modeling of the building process of projects with utility assessment and the main principles of the integrated cost estimation, project management and quality assurance microcomputer based system developed recently. This expert system is based on quick modeling of the building process by use of typical construction technology network diagrams, which can be prepared in advance. The typical network diagrams can be modified according to the spatial conditions of a certain building and to the amount of construction works and materials. For utility assessment a vector of 10 main aspects (criteria) was created with a common measure unit and certain level of importance each. A database of construction processes was created including the aspects for utility assessment. Thus, the model of the building process can be made about 50 times quicker than by current project management systems and it can be used for bidding, project planning and management and utility assessment.

Keywords: Construction technology · Building process · Network analysis · Model · Utility assessment · Project management

1 Introduction

When projects are to be undertaken they have to be efficiently planned at the investor's (owner's), architect's and contractor's sides. In the planning and design stage of the project several specific problems must be solved and many points of view have to be taken into this decision process. A lot of these questions can be efficiently solved by creating of a computer model of the building process of the project. The model must be based on the construction technology and cost analysis and must reflect all technical, technological and economical features. One of the significant questions is the utility of the project, too.

The utility can be defined as the quality or condition of being useful; usefulness. It is a very general concept for the user to be able to evaluate the utility of a building or of a project on the base of this definition. Especially the architect and the investor should have a simple computer method to be able to assess the utility of the project he wants to undertake or design. The utility of the project is a very complex idea. It is a hard work to assess the utility - it is not possible to evaluate the utility only as one magnitude, as for different participants of the building process – the owner (final user), the architect and the contractor of the project the utility of the building or of the project means something very different. Different features of the utility can be defined from the point

A. Kravcov et al. (eds.), *Durability of Critical Infrastructure, Monitoring and Testing*,
Lecture Notes in Mechanical Engineering, DOI 10.1007/978-981-10-3247-9_7

of view of the whole community. Therefore more aspects have to be taken into the method of the utility assessment and a vector of criteria of utility assessment must therefore be created, see e.g. [2]. Naturally, all aspects do not have the same importance, so they must be allotted their weights (importance). It would be very advantageous if all utility assessments aspects could have the same exact measure unit and if they could behave as certain resources, so that they could be calculated mathematically. Then they could be put into a project computer modeling system as resources linked to building constructions and in this way the utility of a whole building or of a project which consists of several building constructions could be evaluated. For modeling of the building or of a project the methodology of construction technology design, see [4], can be very efficiently used. Therefore a lately developed project preparation and management system which includes the database of building constructions with 50 resources each and a big set of typical network diagrams that enable very quick creation of the model of a building or of a project consisting of more facilities, see [3], can be then used for the calculation. With the help of the system the user has the possibility to simulate the proposed composition of constructions and construction processes in the project and the time and resource allocation flow of the building process on a microcomputer even if the topical relevant data about the project in the planning stage are poor. The more precisely the task is determined; the better results can be obtained from the model. The method for the utility assessment of building and projects designed by the author is based on this concept and is briefly described below.

2 Main Utility Assessment Aspects

During the design of the criteria for utility assessment of buildings and project three main areas for the evaluation were stated – the contractor, the investor (owner) and the community. About 15 aspects for all areas were identified and finally 10 main criteria were picked out for the definition of the utility assessment vector. The criteria of the utility assessment vector are as follows:

1 for the contractor:

- realization technology
- environmental impact during the building process
- energy intensity during the construction
- risk during the construction

2 for the investor (owner):

- lifetime and fire resistance
- quality
- operation energy intensity

3 for the community:

- build-up of free area
- noise, traffic, operation emissions
- recycling of used materials

For all aspects a common measure unit was proposed – 1 finyar (fiňár in Czech), abbreviation 1 FIN. It is something like a special sort of currency and in this way all aspects can work as resources in the utility vector assessment calculation. The lesser the amount of finyars will be allotted for a certain criterion the better is the resulting assessment. The importance of all aspects was evaluated by a group of four experts. The aspects were mutually compared one to each other and at every comparison the value 1 was divided into two parts according to the higher of lower importance of the two compared aspects. If the importance of both compared aspects was equal both of them got the value 0.5. If one aspect was more significant than the other, it got the value from 0.51 to 1.0, the other got the value of the complement to 1. The actual values of weights (importance) of all criteria were then calculated as the average of values defined by mentioned four experts.

3 On Modeling the Project

Three main means for quick modeling of a building that consists of constructions or of a building process which consists of construction processes were created. First, a database for the quick modelling of the building is available. This database consists of the main data about all constructions and construction processes at the technological structure of work gangs. It includes main facts about the time standard, productivity of labour, price of the product, number of workers, technological pause and other 20 economical and technological resources (material costs, wages, costs for machines, overheads, average profit, machines, materials, professions) and the utility aspects vector per measure unit of the construction. The values of the utility aspects are given in the interval < 0 FIN/m. u.; 1000 FIN/m. u. > . The neutral value for each aspect is 500 FIN/m. u., the "good" values are lesser than 500 FIN/m. u., the "bad" values are greater than 500 FIN/m. u. The greater the value of FIN/m. u. of the aspect is the worse is the influence of the aspect on the utility. Nowadays, there are more than 500 building constructions and construction processes included in the database, see Figs. 1 and 2.

Second, a set of typical network diagrams of certain types of facilities and their way of erecting based upon the construction technology network diagram method, see [3, 7], was created. The typical network diagram of a building process of a facility as a computer file contains the data about the sequence of the constructions and construction processes, and their linkage. The typical network diagrams can be easily modified according to the spatial structure of the actual building. The volume of production and costs and price of all constructions are included too. They are related to an adequate custom-made measure unit, usually m3 of build-in space or m2 of reconstructed area in case of a reconstruction. The typical construction technology network diagram can be modified according to the spatial structure of the building process by using the 3 main minimum working space indices.

Third, the CONTEC project preparation and management system, see [3], was adapted so that it is capable to calculate and evaluate the utility vector including the weighted values in a certain tabular form.

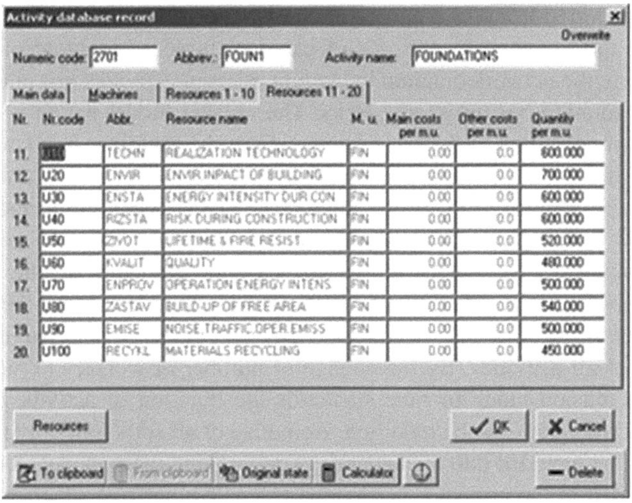

Fig. 1. Part of the listing of the database of construction processes.

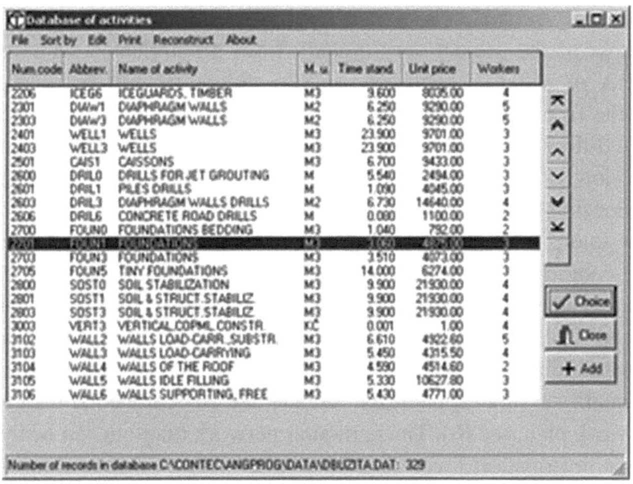

Fig. 2. Chosen activity with utility assessment aspects.

4 Creating the Project Model and Calculating the Utility Vector

When the user wants to simulate the facility or the building process, in the first stage he defines up to 5 custom-made measure units for the building or for the project. Then he calls up the typical network diagram of the certain type of facility, modifies it by stating the actual main working space indices and the computer generates the first draw of the model of the composition of the building, including the time and cost analysis data which are transferred from the database of constructions. Thus, the user can get the first model much quicker than by the use of classical project management systems that require creating the network diagram by adding relevant activities one after another and stating their duration, resources and links. The created model has to be defined with more precision according to the facts known about the building. It is known that 80% of the price and costs is influenced by 25–30% of activities only. Volumes of production of these significant processes have to be stated according to the construction design, prices of the production, labour consumption and resources needed are calculated automatically according to the database of activities. If the exact bill of quantities is available its values can be automatically transferred into the model. After the calculation of the network diagram the user gets the early and late terms of starts and finishes of all activities. By the change of number of workers in the gangs or by changing the tension index of time standards the duration of activities and thus the whole network diagram can be modified. Activities of all sorts (not only those from the database) can be included into the network. After making these models for all buildings that are included in the project it is possible to make a network diagram of the whole project by connecting the partial networks and linking them together with flow links in case of continuous work of specialized work gangs in linked buildings.

Thus, it is not necessary to create the network diagram individually from the very beginning for every project. The expert system enables to build up the model of the building process of the project very quickly from prefabricated sections, typical network diagrams of different facilities, and to define it with more precision easily according to the facts gained from the investor's task. It can be easily updated in case of a change of different conditions. The network can be automatically recalculated from the point of view of keeping the deadline of the project required. The system then selects activities that have to be shortened by adding a certain number of workers or by increasing the intensity of work, while keeping technological rules and all links of the network. The system enables to print the calculated network diagram in different forms (technological standards, bar chart, line-of-balance graph, resource allocation graphs of price, costs and cash flow, labour consumption, need of work force etc.), in Czech, Slovak, English, Italian or Russian. Even in the very first stage of the plan it is possible to create the quality assurance checklist, see [1], the environmental plan, see [5, 8], and the safety-at-work plan, see [6]. The particular network diagram can be then aggregated into the higher information level of technological stages, steps of completion, to the level of facilities and others.

All documents that are gained on the base of the construction technology network diagram can be easily updated according to the actual completion of construction processes on site at a certain term. If there is a delay, the system suggests what measures are to be done to be able to keep the final deadline of the project. At the same time it keeps the technological rules of the building process. This can be visualized in the comparative bar chart, where the updated version of the building process drawn in thicker lines is compared in one document with the planned flow of the process.

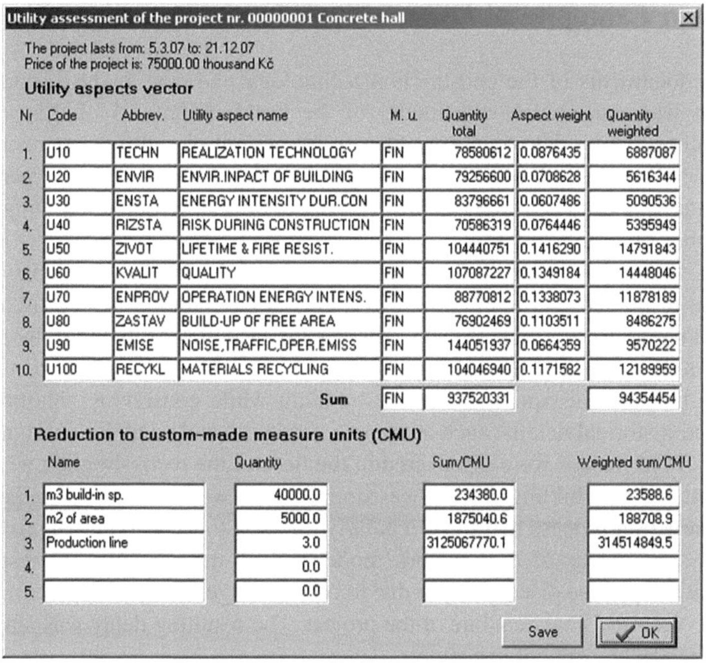

Utility assessment of the project nr. 00000001 Concrete hall

The project lasts from: 5.3.07 to: 21.12.07
Price of the project is: 75000.00 thousand Kč

Utility aspects vector

Nr	Code	Abbrev.	Utility aspect name	M. u.	Quantity total	Aspect weight	Quantity weighted
1.	U10	TECHN	REALIZATION TECHNOLOGY	FIN	78580612	0.0876435	6887087
2.	U20	ENVIR	ENVIR.INPACT OF BUILDING	FIN	79256600	0.0708628	5616344
3.	U30	ENSTA	ENERGY INTENSITY DUR.CON	FIN	83796661	0.0607486	5090536
4.	U40	RIZSTA	RISK DURING CONSTRUCTION	FIN	70586319	0.0764446	5395949
5.	U50	ZIVOT	LIFETIME & FIRE RESIST.	FIN	104440751	0.1416290	14791843
6.	U60	KVALIT	QUALITY	FIN	107087227	0.1349184	14448046
7.	U70	ENPROV	OPERATION ENERGY INTENS.	FIN	88770812	0.1338073	11878189
8.	U80	ZASTAV	BUILD-UP OF FREE AREA	FIN	76902469	0.1103511	8486275
9.	U90	EMISE	NOISE,TRAFFIC,OPER.EMISS	FIN	144051937	0.0664359	9570222
10.	U100	RECYKL	MATERIALS RECYCLING	FIN	104046940	0.1171582	12189959
			Sum	FIN	937520331		94354454

Reduction to custom-made measure units (CMU)

	Name	Quantity	Sum/CMU	Weighted sum/CMU
1.	m3 build-in sp.	40000.0	234380.0	23588.6
2.	m2 of area	5000.0	1875040.6	188708.9
3.	Production line	3.0	3125067770.1	314514849.5
4.		0.0		
5.		0.0		

Save ✓ OK

Fig. 3. Calculation of the utility assessment.

According to the recalculated network diagram the line-of-balance graph (time-space graph) can be automatically drawn on the plotter.

After creating the model of the project the user can transfer the data about the utility aspects for all activities from the database mentioned above. The system then calculates the whole utility vector in a table form, see Fig. 3.

All documents that are gained on the base of the construction technology network diagram can be easily edited and updated according to the actual compositions of constructions. The utility assessment method is completely free for the user. He can define his own utility assessment vector composed of his own aspects, he can identify his own values of weights (importance) of these aspects, he can even design his own database of constructions with allotting of the amount of FIN to every utility aspect in every building construction that is included in the database.

Nowadays, there are 4 types of these databases available – the database for housing and structural engineering, the database for industrial buildings, the database for engineering structures (bridges, roads and infrastructure) and the database for hydraulic structures.

5 Chosen Examples of Use

The main documents of the construction technology and cost analysis created in the mentioned way can contain the model of the building process of the project that includes all necessary data for the building process control and management, resource allocation balancing and utility assessment. The described way of preparation and management was used in some cases of reconstruction of historical buildings and in a large number of significant new projects.

A very interesting example of use of the system for modeling and cost and utility assessment of the reconstruction process of a historical building is the Toscano Palace near the Prague Castle, see Fig. 4. This renaissance palace belongs to the Czech Ministry of Foreign Affairs. The main purpose of reconstruction was to modernize especially the internal equipment of the building while conserving significant architectural and historical details, such as portals, towers, façade and the courtyard. In this case a lot of old frescos were discovered in the flow of the reconstruction which had to be renewed and the building process was stopped for 8 weeks. In that case the model of the reconstruction process had to be regularly updated (once a month approximately.). The main significance of the network model was in the capability of respecting of delays caused by these discoveries or due to other changes and to calculate measures to be done to keep the final deadline of the project. The resulting delay was only 3 weeks compared with the original schedule.

Fig. 4. Toscano Palace in Prague - western façade.

As an example of new significant projects it is worth noting the 20 storey block of flats U Sluncové in Prague, see Fig. 5. It was being built in one of the smallest building sites from the point of view of the area and it was finished in 2009. The system was used for creating of the particular time plan, utility assessment and quality assurance checklist at the start of the project for both the developer and the contractor.

Fig. 5. Block of flats U Sluncové in Prague

6 Conclusions

This paper briefly describes a new design of the methodology of the utility assessment of buildings and projects. It is based on a vector of 10 main aspects (criteria) for utility assessment which was designed with a same measure unit. The aspects were given a certain level of importance each. A database of construction processes was created including the aspects for the utility assessment. On the base of the methodology of the construction technology design a method for modelling of realization of building and projects based on the construction technology network diagram with utility assessment was created and a computer system for this purpose was developed. This system is capable of a quick modelling of a building of a project that is composed from constructions by means of typical construction technology network diagrams. It enables to calculate and tabularly illustrate the resulting utility vector and its sum with regard to the importance of every utility aspect. The methodology is completely free for the user who can define his own values of every utility aspect, even the aspects themselves.

The main documents of the construction technology design created by the help of the expert system can contain the model of the building process of the project that includes all necessary data for the building or reconstruction process control and management. The system enables to create the building process model with appropriate cost assessment and time-cost analysis about 50 times quicker than current project management systems. It is possible to use the documents as a part of a feasibility study, bid, construction technology design and operative plan for the project management of the erection or reconstruction process itself. This model can be updated according to the bill of quantities or cost estimation. Afterwards the quality assurance checklist, the environmental plan and the safety-at-work plan can be automatically created. All documents can be easily updated according to the actual completion on site at a certain term. If there is a delay, the system suggests what measures are to be done in the future to be able to keep the final deadline of the project. All documents can be automatically translated to Czech, Slovak, English, Italian and Russian.

The designed methodology of the utility assessment and modeling of the building process of facilities and projects can be used especially in the phase of planning the projects by owners, designers and architects for utility assessment and much more – for feasibility studies, planning, obtaining bank loans to finance the projects and last but not least in project management itself.

The described system can be used on IBM PC compatible computers under Windows 9x, ME, NT, 2000, XP, Vista and Windows 7, 8 and 10 operational systems.

References

1. Gašparík, J.: Manažérstvo kvality v stavebníctve (Quality Management in the Building Industry). Vydavateľstvo Jaga group, v. o. s., Bratislava (1999)
2. Jain, A.: Project thinking – an essential ingredient for value enhancement. In: Proceedings of the 1st ICEC & IPMA Global Congress on Project Management, Ljubljana. ZPM Slovensko združenje za projektni managment Ljubljana, CD – paper # PL – 02, Slovenia (2006)
3. Jarský, Č.: Automatizovaná příprava a řízení realizace staveb (Automated Preparation and Management of Realization of Structures). CONTEC Kralupy n. Vlt., Kralupy nad Vltavou (2000)
4. Jarský, Č., Musil, F., et al.: Příprava a realizace staveb (Preparation and Realization of Structures). Akademické nakladatelství CERM s. r. o., Brno (2003)
5. Jarský, Č.: K tvorbě environmentálních plánů v přípravě staveb (On Creating Environmental Plans in Building Preparation). Časopis Stavebnictví 2009/02, pp. 62–65, Czech Republic (2009)
6. Jarský, Č., Gacho, P.: Počítačové zpracování agendy bezpečnosti a ochrany zdraví (Computer Elaboration of the Safety-at-work Agenda). Časopis Stavebnictví 2010/02, pp. 60–62, Czech Republic (2010)
7. Jarský, Č.: Modelling of the building process in construction planning and management using I. T. In: Proceedings of the 2010 – 27th International Symposium on Automation and Robotics in Construction (ISARC), Bratislava, Slovakia. Tribun EU Brno, Czech Republic (2010)
8. Makýš, P.: The automation of construction scheduling considering variability of working environment. In: Proceedings of the 2010 – 27th International Symposium on Automation and Robotics in Construction (ISARC), Bratislava, Slovakia. Tribun EU Brno, Czech Republic (2010)

Design Tool for Static Design and Evaluation of Steel, Wood, Concrete and Masonry Members in Buildings

Jan Holub[1(✉)] and Pavel Maňas[2]

[1] 151. Engineer Battalion, Bechyně 391 65, Czech Republic
holubj@army.cz
[2] Faculty of Military Technology, University of Defence,
Kounicova 65, 662 10 Brno, Czech Republic
pavel.manas@unob.cz

Abstract. This paper is focused on the creation of the series of design tools for design and assessment of individual construction elements made of steel, wood, concrete and masonry. These design tools are created as the individual files in Mathcad and are accompanied by the flowchart with computing algorithm. All proposals are performed in accordance with the EU standards.

Keywords: Eurocodes · Mathcad · Structural design

1 Introduction

Engineers in the Army of the Czech Republic have to deal with various task during their career in the army. After finishing the University of Defence each officer is assigned to different unit and started be more specialized in his specific task. Knowledge from the University of Defence is not used in a full range and some abilities begin to fade. One of these abilities is structural calculations, which are forgotten when is not used regularly. However, sometimes there are not enough of specialized engineer officers, for every task, which could appear, for example in foreign missions. At this time engineer officer with different specialization have to be used to deal with specific problem like structural calculation in this case. For these situations was design tools created, to help engineer officer deal with basic structural calculation. This paper has two parts. First is design tool created in the Mathcad, which could be used for calculations on the computer. Second are flowcharts showing computing algorithms to change the algorithms or to be used for manual calculation as a guide.

2 Design Tools

Currently, due to the great development of computer technology, there is a possibility of performing structural calculations in designing of the structures by using of the various types of computing software. Nowadays on the market there are many programs that deal with this topic, but they are mostly commercial programs. These programs are

A. Kravcov et al. (eds.), *Durability of Critical Infrastructure, Monitoring and Testing*,
Lecture Notes in Mechanical Engineering, DOI 10.1007/978-981-10-3247-9_8

often complex and it is not possible to intervene and change their source code, which means that possible changes are very limited or impossible.

For this reason the design tools have started be developed. It deals with creating a series of design tools that could be used within the Department of Engineer Technologies. Department of Engineer Technologies within the Army of the Czech Republic has to follow valid standards applied in the Czech Republic. Because of the membership of the Czech Republic in the EU implemented standards are based on the Eurocodes [1–6].

Also the other armies are by law obliged to follow legal standards that guarantee a certain norm when building structures for their trouble-free using. These standards are also important in foreign missions where we expect rotations of different countries units. By using Eurocodes the fact that the buildings will be legally in accordance with all EU member states is achieved. Due to the growing number of members of the EU the importance of using Eurocodes is increasing.

For this reason the design tools are created in accordance with the Eurocodes.

The actual design tools were originally created in Mathcad 15, however due to the development it has been subsequently transferred into the program PTC Mathcad Prime 3, which includes the license for the program Mathcad 15.

As was mentioned in the introduction, there are now different variations of computing and design programs available on the market. We can meet with tools for creating conceptual designs and detailed designs for entire buildings. The creation of detailed proposals are mostly programs using the FEM method. These programs are characterized by high complexity and the possibility of designing whole buildings. However, these programs are also quite complex to use and for the user is needed to be familiar with the using of these programs. That could be time-consuming and in the case of special courses quite costly. These commercial programs are also often characterized by high acquisition price. These type of the design tool is not important for occasional user in the form of engineer officer. So was decided that the result of this work will be conceptual design tools.

Therefore it was decided that the created design tools must be created in one of the programs that has a purchased license within the University of the Defence and students could commonly meet with it. For the reason that students, from whom will became engineer officers, should be able to use this programs without special course. The program should be characterized by a user-friendly environment, simplicity and clarity of the programming code, so end users has a clear and possibly changeable tool in front of him that could further use and modify as required.

Programs like Maple, Matlab and Scilab were not chosen, not because of their computing possibilities, but because of their complex programming code that could be for many users difficult and too complex. Possibility of creating a graphical interface for these programs has been evaluated as tedious and also for the user who is not familiar with these programs has limited possibility of changes in the individual design tools. Microsoft Excel has to be excluded, despite its development and opportunities within the basic programming and calculations, because of the difficult orientation in the complex calculations it would be very difficult to change the design tool for the end user.

From all the above listed reasons, and in comparison with other programs that are commonly used at the University of the Defence the chosen program was Mathcad. One of the main advantages of this program belong for example a natural form of mathematical notation, which is understandable to ordinary users at first sight. The program is able to use numerical and symbolical calculations, systems of equations, creating graphs and even more. It also contains the necessary computing and programming tools to create the necessary computing algorithm. Its other advantage is the ease of editing and the ability to connect the various images, text and the ability to import and export data in various formats. Together with the use of floating license it was chosen as the appropriate program for creating of the design tools.

Design tools for steel members. Calculations for design tools for steel members are developed in accordance with EN 1993-1-1. Members could be loaded by bending, tension, compression and buckling compression. These design tools have the common part, which defines the properties of the steel.

Design tools are created for cross-section IPE, IPN, HEB, UPE and hollow tubes. To select a specific design tool, the user have to select the specific file of design tool for the given cross-section. Cross-sections of type IPE, IPN and HEB are collectively referred to as cross-section of I type.

Design tools for timber members. Calculations for design tools for timber members are developed in accordance with EN 1995-1-1. Members could be loaded by bending, tension, compression and buckling compression. These design tools have the common part, which defines the properties of the timber according to EN 338.

Example how the design tool appears is shown on Fig. 1. Most of the design tools have similar appearance. From this reason only one example is shown.

Design tools for the masonry members. Calculations for design tools for masonry members are developed in accordance with EN 1996-1-1. Members could be loaded by continuous force and concentrated force.

Design tools for the concrete members. Calculations for design tools for concrete members are developed in accordance with EN 1991-1-1. Members could be loaded by bending, tension, compression and buckling compression. These design tools have the common part, which defines the properties of the concrete (Fig. 2).

Inputs of the design tools. Data input is made in the most intuitive way as was possible. At the beginning of the first page there are given basic information about the particular form of the design tool. For what type of the element, loading and material is tool designed and is specified by which standard calculations are performed. Also the static schema, which is used for illustrative picture for the user and displays the main variables entered by the user to the calculation, is included.

Design tool for wooden columns and poles

Design tool for designing wooden poles and columns. This tool includes the effect of losing of stability during buckling. Utility is able to calculate the element of circular or rectangular cross-section of different boundary conditions and loaded by one axial force.
Calculations are performed according to Eurocode 5: Design of timber structures.

Input variables:

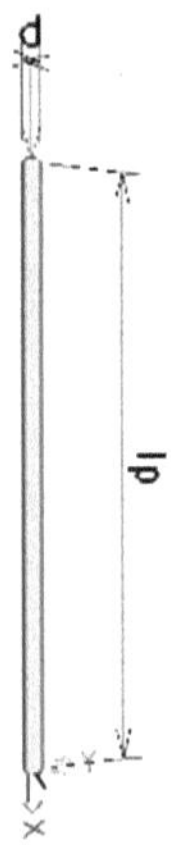

$N_{c\_k0} := 250\ kN$ ... characteristic axial force parallel to the grain

$a := 240 \cdot mm$ ...distance of the element

$b := 240 \cdot mm$ ...width of the element

$d := 0 \cdot mm$ ...diameter of the element

$l_p := 4 \cdot m$ ...lenght of the element

$action := 1$ Partial safety factors for actions in building structures for persistent and transient design situations.
Type:
1 for permanent
2 for variable
3 for exceptional

$timber_v := 1$ Type:
1 for softwood SI
2 for softwood SII
3 for hardwood SI

$load\_duration_v := 1$ Select the predominant term of load - type:
1 for permanent - more than 10 years (e.g. self weight)
2 for long-term - 6 months to 10 years (e.g. storage)
3 for medium-term - 1 week to 6 months (e.g. imposed load)
4 for short-term - less than one week (e.g. snow and wind)
5 for instantaneous - e.g. accidental load

$service\_class := 1$
Type:
1 for Service class 1: is characterized by a moisture content in the materials corresponding to a temperature of 20 °C and the relative humidity of the surrounding air only exceeding 65 % for a few weeks per year
2 for Service class 2: is characterized by a moisture content in the materials corresponding to a temperature of 20 °C and the relative humidity of the surrounding air only exceeding 85 % for a few weeks per year
3 for Service class 3: climatic conditions leading to higher moisture contents than in service class 2

Page 1 of 2

University of Defence
Department of Engineer Technologies

Developed by: OR-3 Jan Holub
jan.holub@unob.cz
www.unob.cz

Fig. 1. Example of design tool page 1 of 2

Data input is made in two different ways. First is input of the necessary values for the calculation. Individual variables are always accompanied by explanatory to enlarge clarity. User due to these explanatory has an overview of what each variable represents and can easily fill in input data. Second is the selection of predefined choices. For

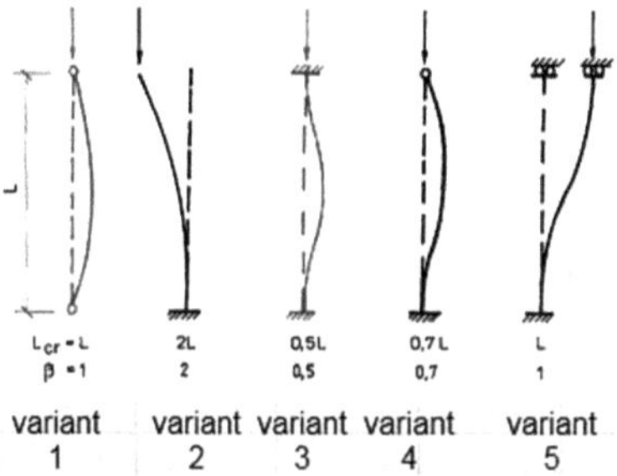

choose one of the boundary conditions - type:
1 for variant 1
2 for variant 2
3 for variant 3
4 for variant 4
5 for variant 5

$boundary\_condition := 1$

Properties of timber:

Calculation:

The results:

The variables in the calculation:

$A_{nc0} = 0.058\ m^2$ ... cross-section area
$k_{mod} = 0.6$ ... coeficient of load duration and moisture (service class)
$\gamma_N = 1.2$ ... partial factor

Variables affecting the stability of the element:

$l_{crit} = 4\ m$ ... critical lenght of element - buckling
$i_{min} = 0.069\ m$ ... radius of gyration
$\lambda_i = 57.73$ ... slenderness ratio
$\sigma_{c\_crit} = 19.841\ MPa$ ... maximum design compressible stress
$\lambda_{rel} = 1.004$ ... relative slenderness
$k = 1.054$
$k_c = 0.726$ ... buckling factor
$\sigma_{c\_0\_d} = 5.208\ MPa$ ... design compresion stresses
$f_{c\_0\_d} = 8.276\ MPa$ ... design compresion strenght

Assessment on the pressure:

$assessment\_pressure =$ "suitable"
$use\_of\_the\_element = 86.629\ 1\%$
... percentage of used loading capacity of axially loaded column with buckling

University of Defence
Department of Engineer Technologies

Page 2 of 3
Developed by: OR-3 Jan Holub
jan.holub@unob.cz
www.unob.cz

Fig. 2. Example of design tool page 2 of 2

increasing of comfort for users design tools includes needed basic database for calculations from which user can choose by typing a right specific value of predefined choices. For the reason that Mathcad Prime 3 compared to the version Mathcad 15 no longer has the "List Box" function, when the user only click to select the required choice of predefined variables, now user have to make the choice by typing the value to the variable according to the explanatory that assigns a specific value, which would be used in the calculations.

Computing algorithm. Part of the algorithm, where the calculations are done and where the input variables of predefined choices are defined, is hidden by the function "Area", so whole document is clearer. On the other hand, the user has the possibility to open this hidden area and possibly change some of the values or formulas. By this is achieved clarity and also gives the ability for the user to change some predefined coefficients, or formulas according to his specific needs.

Outputs of the design tools. Basic output is the assessment at the end of the document, whether the load-carrying capacity of the element is exceeded or not. It is described how many percent of the cross-section of the element is used. This information is rather tentative for the need of economic sizing and is based on a simple assessment of load of the element to bearing capacity of the element.

Other possibilities are graphical outputs that help the user to get a better overview of the load and the use of the element. These include a graph displaying the progress of the bending moment on the beam or assessment of the bearing capacity of the compressed element made of reinforced concrete using interaction diagram that shows the boundaries of the inner limit forces. Evaluation of bearing capacity is performed by the user like assessment, whether a given loading condition, which is showed by the blue cross, is located within the limit of the internal forces and significant points, which are showed by red lines and red crosses.

3 Flowcharts

On each of design tools is made flowchart. These flowcharts could help to end user understood computing algorithm. Meaning of it is, when user need to change some part of computing algorithm, for what exactly he needs. In military use it could be some kind of factor, which is not considered in Eurocodes and have to be changed, for example from safety reasons.

Another reason for creating the flowcharts is that engineer officer, who is long time from his studies could use flowcharts to revive his ability calculate basic structural elements. User could follow the flowchart and step by step be led through calculation. These flowcharts could be also used in the education as aid.

Flowchart itself is made in the version how the computing algorithm in Mathcad Prime 3 is done (Figs. 3 and 4).

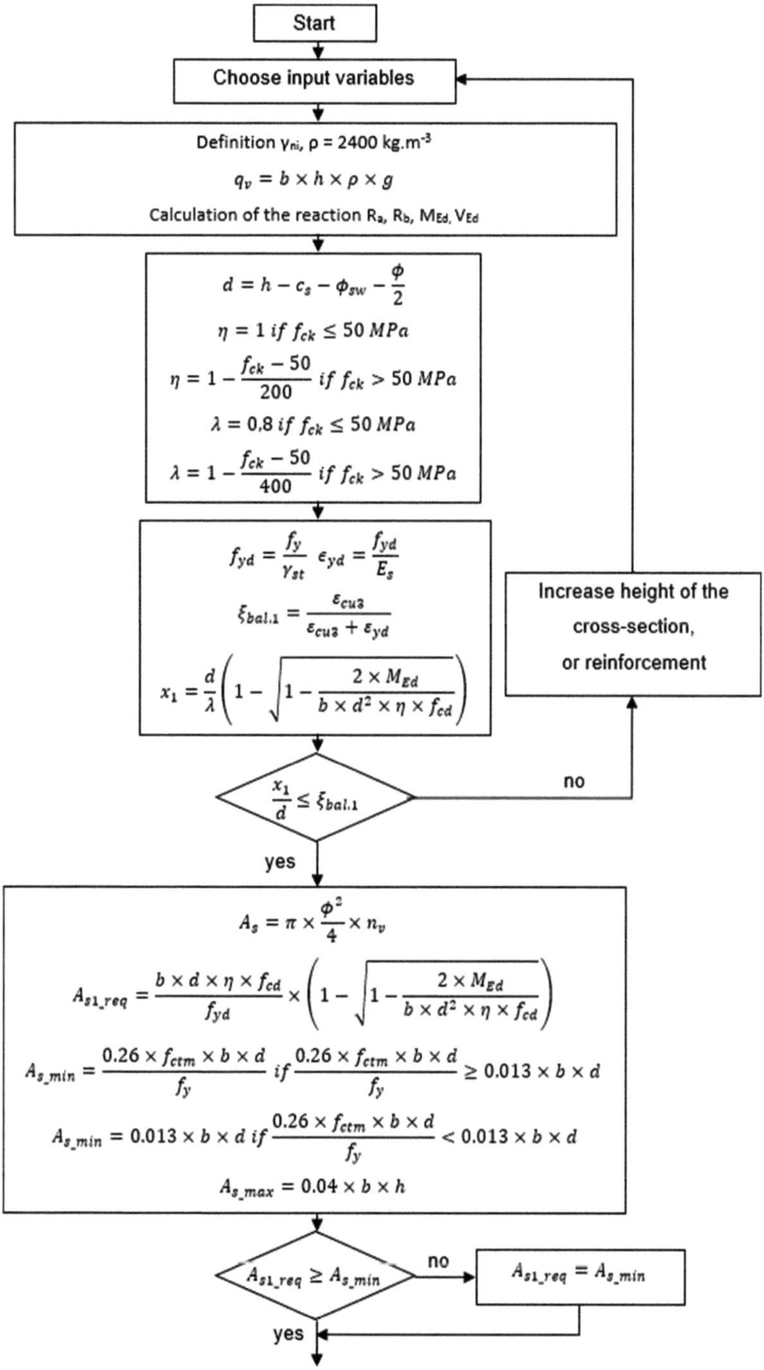

Fig. 3. Example of the flowchart part 1

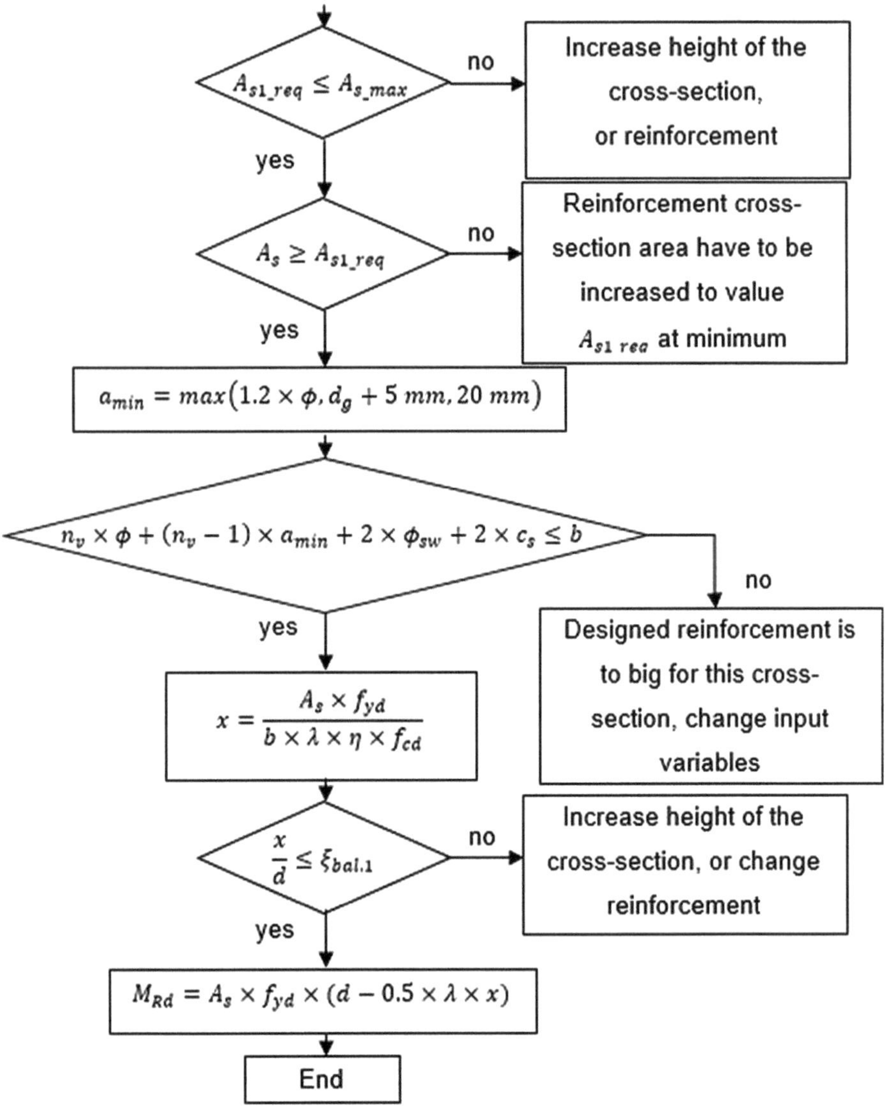

Fig. 4. Example of flowchart part 2

4 Finished Design Tools

So far these individual design tools including basic load were created:

For wood:

- tension – perpendicular to the fibres, parallel to the grain
- compression – perpendicular to the fibres, parallel to the grain with the loss of stability due to buckling

- bend – without the loss of stability and with the loss of stability due to lateral buckling

For steel:

- tension – for profiles of type I, UPE, and the circular cross section
- compression – without the loss of stability and with the loss of stability due to buckling for profiles of type I, UPE, and the circular cross section
- bend – without loss of stability and with the loss of stability due to lateral buckling for profiles of type I

For masonry:

- pressure – continuous and concentrated load

For concrete:

- bend – an assessment of the bending and shear, the assessment of the bending and shear according to EN 1992-1-1
- compression – the assessment of the pressure according to EN 1992-1-1

For these design tools were also created flow charts.

These design tools has proven to be beneficial for use in the classroom. Students can use it to control their own calculations, giving them the opportunity to study algorithms. With it they can also deal with the individual steps in the calculation of more complex examples and student can handle the task quickly, focusing on specific steps of the assessment in the design.

The systematic development can achieve usability in various activities of the Army of the Czech Republic, mainly as a support of the information portal of engineer army and in the concept of the Reach-back. These design tools will be placed to the information portal of engineer army of the Czech Republic, where they should be available under the administration of the Department of the Engineer Technologies of the University of Defence.

5 Conclusion

For the usage Army of the Czech Republic and its engineers these design tools for designing the static structures of certain elements of wood, steel, masonry and concrete in the basic states of the stress were created. These design tools are characterized by ease and rapidity of the using, which could be essential condition for use by all engineer officers in the army. These factors could help engineer officer to do his task with less time used for reviving of his knowledge about structural design. Printed flowcharts could help him in case that he is in a situation, where is not possible to use computer technology.

For the assessment of complex structures, it is necessary to use commercial programs. These professional programs include calculation algorithms for almost all load cases. They were created by teams of experts and used by many users, their calculation algorithm has been tested and no errors should be in it.

Despite some of its sub-cons can be used in the concept of the Reach-Back, where a remote user or a member of Department of the Engineer Technologies could use this program to design the individual elements. For this purpose appearance of the design tools were developed into a form suitable for printing a document. Annexed flowcharts could help user to understand calculation algorithm, so he should be able to change algorithm for his own purpose.

At least these design tools could be used as part of the education process and students could use it and change it for what they need.

Acknowledgment. The work presented in this article has been supported by the Czech Republic Ministry of Defence - K 201 Department development program "Development of technologies in weapons construction, ammunition, weapons instrumentation, materials engineering and military infrastructure".

References

1. EN 1990: Eurocode - Basis of structural design
2. EN 1991-1-1: Eurocode 1: Actions on structures - Part 1-1: General actions - Densities, self - weight, imposed loads for buildings
3. EN 1992 - 1-1: Eurocode 2: Design of concrete structures -Part 1 - 1: General rules and rules for buildings
4. EN 1993 - 1-1: Eurocode 3: Design of steel structures - Part 1 - 1: General rules and rules for buildings
5. EN 1995 – 1 -1: Eurocode 5: Design of timber structures -Part 1-1: General - Common rules and rules for buildings
6. EN 1996 -1-1: Eurocode 6: Design of masonry structures -Part1-1: General-Common rules and rules for buildings

Measurement of Residual Stresses in Alloys Using Broadband Ultrasonic Structuroscopy

Alexander A. Karabutov[1,2(✉)], Elena B. Cherepetskaya[1], Alexander Kravtsov[3], Vladimir A. Makarov[2], Elena A. Mironova[1,2], Dmitry V. Morozov[1], and Pavel Svoboda[4]

[1] Laser Ultrasonic Nondestructive Testing Lab, The National University of Science and Technology "MISiS", Leninsky pr.6, Moscow 119991, Russia
aak@optoacoustic.ru, {echerepetskaya, moroz_off_123}@mail.ru, lenamironova71@yandex.ru
[2] International Laser Center of M.V.Lomonosov Moscow State University, Leninskie gory, 1, Moscow 119991, Russia
vamakarov@phys.msu.ru
[3] Czech Technical University in Prague, Thákurova 7/2077, 166 29 Prague 6 Dejvice, Czech Republic
kravtale@fsv.cvut.cz
[4] Department of Construction Technology, Faculty of Civil Engineering, Czech Technical University in Prague, Prague, Czech Republic
pavel.svoboda@fsv.cvut.cz

Abstract. Broadband ultrasonic spectroscopy used to measure residual stresses in samples of high-temperature alloys. It is shown that the acoustoelasticity method and laser ultrasonic spectroscopy can be used to determine the depth distribution of residual stresses. Measurements were carried out on test samples of heat-resistant nickel-base and titanium-base alloys with specified distribution of residual stresses after vacuum heat treatment, vibratory stress relief or shot peening. The results were compared with the residual stresses determined from changes in the deformation upon the removal of layers from the sample by electrochemical etching.

Keywords: Residual stress · Alloy · Broadband ultrasonic structuroscopy · Contour · Neutron diffraction · Relative variation of acoustic impedance

1 Introduction

Residual stresses occur in many manufactured structures and components. The engineering properties of materials and structural components, notably fatigue life, distortion, dimensional stability distortion, dimensional stability, corrosion resistance, and brittle fracture can be considerably influenced by residual stresses [1]. Any method of residual stress testing are classified as either destructive or semi destructive or non destructive [2]. Sectioning [3], contour [4, 5], hole-drilling [3] are destructive techniques. The resulting deformation due to residual stress redistribution is measured in these methods. Non destructive methods include X-ray [6] or neutron diffraction [6, 7], ultrasonic methods [8] and magnetic methods [9].

A. Kravcov et al. (eds.), *Durability of Critical Infrastructure, Monitoring and Testing*, Lecture Notes in Mechanical Engineering, DOI 10.1007/978-981-10-3247-9_9

One of the promising directions in the development of nondestructive techniques for residual stresses measurement is the application of ultrasound. Ultrasonic stress measurement techniques are based on the acoustic-elasticity effect, according to which the velocity of elastic wave propagation in solids is dependent on the mechanical stress. Mechanical stresses induce local changes in ultrasound velocity and density of the medium. The medium becomes acoustically heterogeneous. As a result, ultrasonic waves are scattered by arising heterogeneities. The scattered waves carry information about the magnitude and distribution of the mechanical stresses.

In this study we used broadband optoacoustic spectroscopy to measure residual stresses in samples of high-temperature alloys; the results were compared with the residual stresses determined from changes in the deformation upon the removal of layers from the sample by electrochemical etching.

2 Methods

The relationship between the ultrasonic reflection coefficient and the depth distribution of residual stresses can be found with the aid of a layered model of the medium whose acoustic impedance Z smoothly changes from one layer to another ($Z = \rho V$, where ρ is the density of material and V is the velocity of longitudinal acoustic waves).

Assuming that the density of the medium is constant, the dependence of the velocity of longitudinal ultrasonic waves on depth x can be written as follows:

$$V(x) = V_0(1+\alpha(x)) \quad \Leftrightarrow \quad \alpha(x) = \frac{V(x) - V_0}{V_0}, \tag{1}$$

where $\alpha(x) << 1$.

Let the thickness of layers be small enough to neglect changes in velocity $V(x)$ and $\alpha(x)$ within each layer. Also, let only single reflections from the boundaries between layers be taken into account. This is reasonable because changes in the acoustic impedance do not exceed 1%–2% and the contributions of re-reflections are small (beyond the accuracy of measurements).

The acoustic reflection coefficient for the i-th layer is described by the formula:

$$R(x_i) = \frac{\rho V^i - \rho V^{i-1}}{\rho V_i + \rho V^{i-1}} \approx \frac{V^i - V^{i-1}}{2V_0} = \frac{1}{2}(\alpha(x_i) - \alpha(x_{i-1})). \tag{2}$$

Hence

$$\sum_{i=1}^{k} R(x_i) = \frac{V^k - V_0}{2V_0} = \frac{\alpha(x_k)}{2} \quad \Leftrightarrow \quad \alpha(x_k) = 2\sum_{i=1}^{k} R(x_k). \tag{3}$$

In the extreme case of infinitely thin layers, proceeding from summation to integration and taking acoustoelasticity relations into account, we have the following expression for the principal stresses:

$$\begin{cases} \sigma_{33}(x_1) = A_1 \times \int\limits_0^{x_1} R(x)dx \\ \sigma_{11}(x_1) + \sigma_{22}(x_1) = B_1 \times \int\limits_0^{x_1} R(x)dx \end{cases} \tag{4}$$

To determine the residual stress depth profile, the ultrasound backscattering coefficient should be measured and integrated. Constants A_1 and B_1 are determined using a calibration procedure.

3 Experimental Results

In this study, a precision instrument was used to measure the velocity of pulsed laser-generated longitudinal acoustic waves; this significantly improved the accuracy of time-of-flight measurements of sound speed and consequently residual stresses. A laser optical-acoustic transducer excited powerful short ultrasonic pulses with pressure amplitude up to 0.1 MPa and duration of 100 ns and received acoustic signals scattered by heterogeneities in samples. The acoustic signals were transformed into electric signals to be directed to a digital oscilloscope and a PC for processing.

The transducer received optoacoustic signals reflected from the surface of the sample and scattered by heterogeneities caused by residual stresses; it is the scattered signals that provide information on the distribution of acoustic impedance and therefore

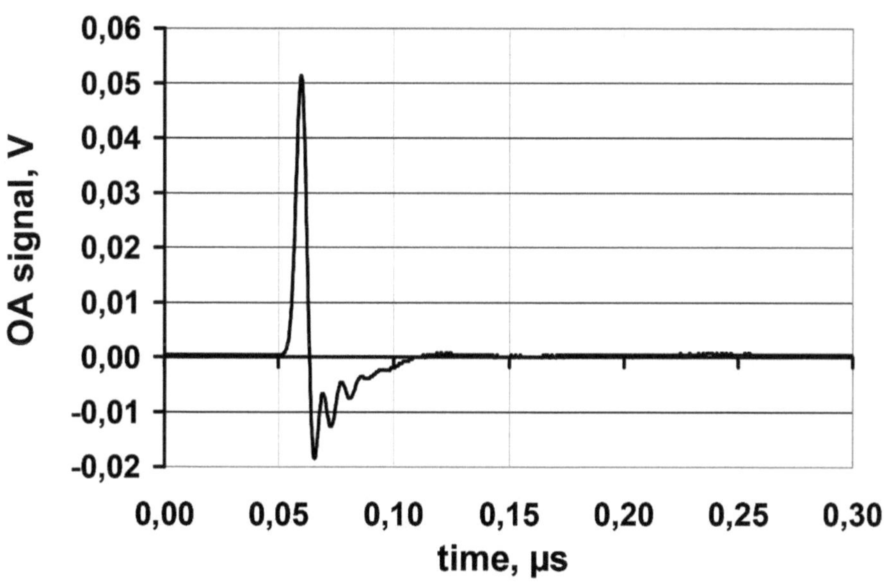

Fig. 1. Typical waveform of detected ultrasonic signal

the residual stresses. A typical waveform is shown in Fig. 1. The oscillating part of the recording carries information on residual stress distribution.

The full-waveform recording shown in Fig. 1 is the convolution of the temporal shape of laser pulse intensity with the transfer function of the transducer and the response function of the medium heterogeneity. To isolate the desired normalized signal, i.e. the response function of the medium, the full-waveform recording was deconvolved with the reference signal obtained on a sample with no residual stresses. The distribution of acoustic impedance Z with depth x was thus determined. Time-to-depth conversion was performed as follows: time was multiplied by half the ultrasound speed in the medium because the arrival time of the signal from a layer is the time that ultrasound takes to travel to the layer and back.

Measurements were carried out in no less than 10 points on the surface of the samples; all recordings were processed as described above. The acoustic impedance measurements were averaged for each depth point. Changes in the averaged acoustic impedance values were converted into the absolute values of stresses after the calibration procedure.

Measurements were carried out on test samples of heat-resistant nickel-base alloys with specified distribution of residual stresses after vacuum heat treatment, vibratory stress relief or shot peening.

When performing measurements by the acoustoelasticity method, first a normalized signal from the surface of a test sample was recorded after heat treatment; the sample had minimum residual stresses at minimum depths. This signal was chosen as a reference one. Measurements on other test samples were compared with the reference level using the procedure described above. The results of the measurement of relative

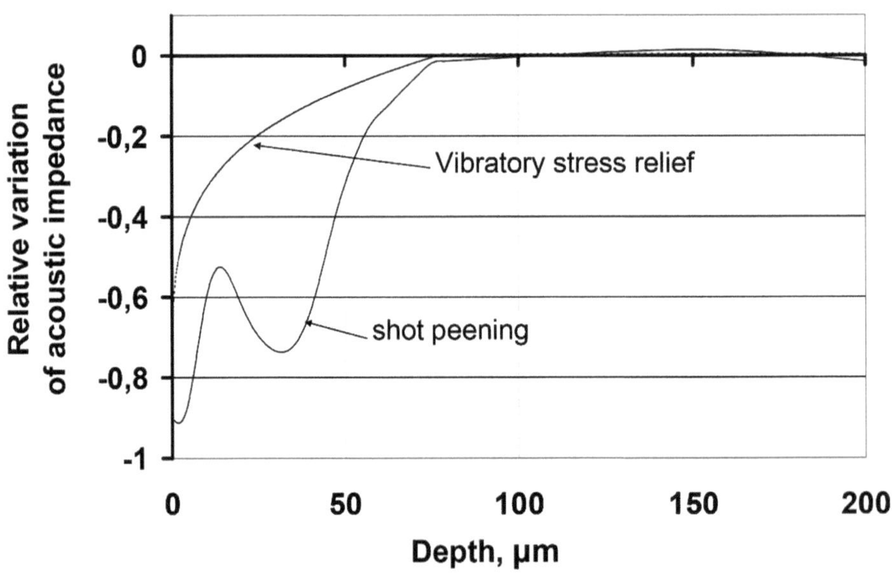

Fig. 2. Heat-resistant alloy samples: relative variation of acoustic impedance versus depth curves (acoustoelasticity method)

depth-dependent variations in acoustic impedance after different types of treatment are presented in Fig. 2.

The results of the acoustoelasticity method were compared with the results of the absolute destructive method for determining residual stresses. In the latter case, residual stresses were determined from changes in deformation due to the removal of layers by electrochemical etching. Figure 3 shows the residual stresses in heat-resistant nickel alloy samples undergone the removal of layers 1 μm thick. It is clear from Figs. 2 and 3 that the acoustic impedance vs. depth curves are similar to the residual stress vs. depth curves.

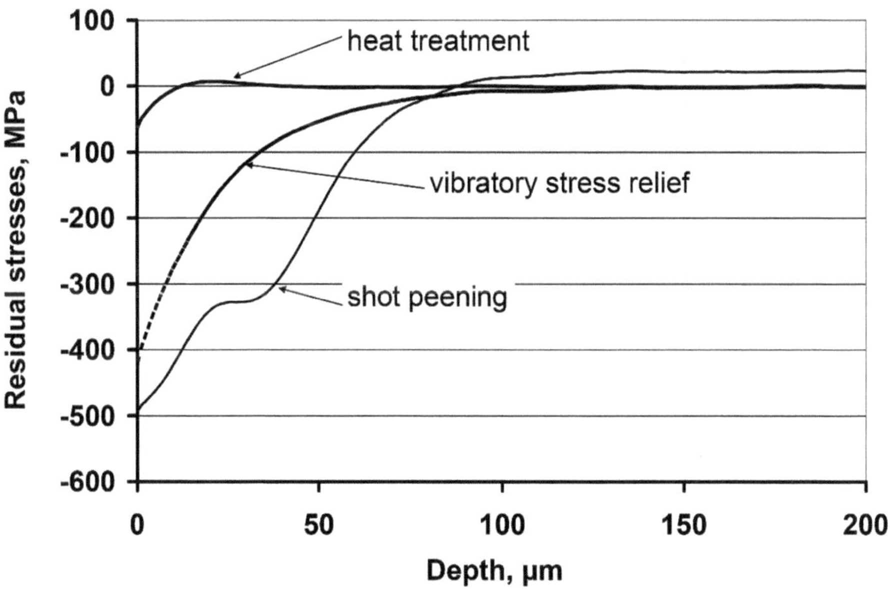

Fig. 3. Specified diagrams of surface residual stresses in heat-resistant nickel alloy samples

The calibration factor was determined as the ratio of the residual stress value to the acoustic impedance value at reference depth points. The calculation was based on 80 points (from 20 μm to 100 μm) since the measurements by the destructive method were carried out with a step of 1 μm. The averaged value was accepted as the calibration factor, which was equal to 580 MPa. The calibration factor for any alloy can be obtained in such a way.

The results of the calibration after vibratory stress relief are shown in Fig. 4. The correlation coefficient for the curves shown in Fig. 4 is 0.983.

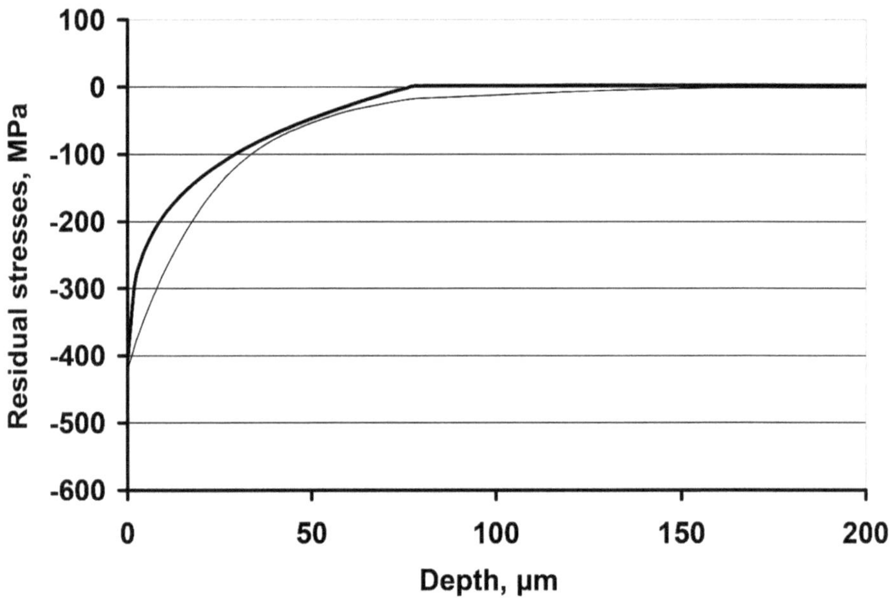

Fig. 4. Comparison of measurements on alloy samples by acoustoelasticity method (thick line) and destructive method (thin line) after vibratory stress relief

4 Conclusions

It is shown that the acoustoelasticity method can be used to determine the depth distribution of residual stresses. A calibration procedure for measurements by the non-destructive acoustoelasticity method has been developed; the method using broadband ultrasonic pulses determines the residual stresses from variations in acoustic impedance.

The calibration procedure based on the use of test samples with a specified distribution of residual stresses can be possibly applied to control the residual stresses in welded joints.

Acknowledgements. The work was carried out with financial support in a part from the Ministry of Education and Science of the Russian Federation in the framework of Increase Competitiveness Program of NUST « MISiS » (№ K1-2015-025).

References

1. Dunn, B.D. (ed.): Metallurgical Assessment of Spacecraft Parts and Materials II. Wiley, New York (1996)
2. Bahadur, A., Kumar, B.R., Kumar, A.S., Sarkar, G.G., Rao, J.S.: Development and comparison of residual stress measure on welds by various methods. Mater. Sci. Technol. **20**, 261–269 (2004)

3. Venkitakrishnan, P.V., Philip, J., Krishnamurthy, R.: An assessment of stresses in thin walled welded tubes through hole drilling and sectioning methods. J. Mater. Process. Technol. **185**, 228–232 (2007)
4. Hosseinzadeh, F., Bouchard, P.: Mapping multiple components of the residual stress tensor in a large P91 Steel Pipe Girth Weld Using a single contour cut. Exp. Mech. **53**(2), 171–181 (2012). doi:10.1007/s11340-012-9627-z
5. Zhang, Y., Pratihar, S., Fitzpatrick, M.E., Edwards, L.: Residual stress mapping in welds using the contour method. Mater. Sci. Forum **490**(491), 294–299 (2005)
6. Zhang, Y., Ganguly, S., Stelmukh, V., Fitzpatrick, M.E., Edwards, L.: Validation of the contour method of residual stress measurement in a MIG 2024 weld by neutron and synchrotron X-ray diffraction. J. Neutron Res. **11**(4), 181–185 (2003)
7. Kartal, M., Turski, M., Johnson, G., Fitzpatrick, M.E., Gungor, S., Withers, P.J.: Edwards L Residual stress measurements in single and multi-pass groove weld specimens using neutron diffraction and the contour method. Mat. Sci. Forum **524–525**, 671–676 (2006)
8. Tanala, E., Bourse, G., Fremiot, M., De Belleval, J.F.: Determination of near surface residual stresses on welded joints using ultrasonic methods. NDT E Int. **28**, 83–88 (1995)
9. Roskosz, M., Bieniek, M.: Evaluation of residual stress in ferromagnetic steels based on residual magnetic field measurements. NDT&E International **45**, 55–62 (2012)

Mathematical Formulation of Innovation of Production Process

Popenková Miloslava(✉)

Faculty of Civil Engineering, Department of Construction Technology, Czech Technical University in Prague, Prague, Czech Republic
miloslava.popenkova@fsv.cvut.cz

Abstract. The article is a proposal of development of anhydrite screed and subsequently carrying out samples of examination of anhydrite screed without fibres and with different kinds of fibres. Furthermore, there are mentioned determination and comparison of mechanical as well as physical characteristics of those different samples. There have been performed calculations of critical point of proliferation of breach by means of fracture energy and cross-cutting module for validation of differences between proliferation breaches of anhydrite fry with and without fibres. Based on the comparison of both results there have been determined different criteria for proliferation of breach for firmness in bending tension and for linear fracture mechanics.

Keywords: Anhydrite screed · Breach · Fracture energy

1 Development of the Process

Monitoring individual elements and their interrelation (input check, takedown check), compare their real quality parameters with standard parameters (output check) and their evaluation should lead to proposal of precautions (especially repeatable defects). This proposal of precautions has to lead to process improvement. During this phase it is necessary to:

- Identify defect – in spite of observance the technological requirements there are cracks in the screed,
- Analyze defect in the parts of production process and during their relation – while object changes into product (anhydrite screed) the screed dries up and cracks create (cracks could cause draught, projecting object – corner, column..)
- Work out proposal of precautions which leads to defect elimination – proposal for using fibres into mixture
- Realize proposal of precautions, whereas development of production process have to go out of elements which are divided into following parts:
  - Input elements – contain information about working object and product (including specification of detected defects) – anhydrite mixture + different kinds of fibres,

A. Kravcov et al. (eds.), *Durability of Critical Infrastructure, Monitoring and Testing*, Lecture Notes in Mechanical Engineering, DOI 10.1007/978-981-10-3247-9_10

- Created elements – contain information about transformation of working object into product, progress of transformation and solution – uniform placement of fibres into anhydrite mixture
- Elements of production – contain results of known production processes analyzes, theoretical branches and their results, practical knowledge, outer influences, …
- development of production process results requirements – minimization of cracks creation

– Monitor course of developed production process with focus on developed elements and relations
– Interpret the proposal of precaution – see Fig. 1.

Fig. 1. Free fall concrete mixer, mixture with polypropylene fibres

2 Specimens for Tests of Anhydrite Screeds

Characterized procedure of development of production process is presented by practical example. New progress enables minimize or eliminate risks which are caused during production process [2, 3] (Fig. 2).

There is proposed development of transformation of working object into product in order to minimize defects caused during realization of anhydrite screeds – i.e. change has been made by new properties of working object (improving its properties). This change goes out of requirement of minimization of the cracks creation during anhydrite screeds realization – during drying phase and during loading phase. There has been made following mixtures in conjunction with TBG Metrostav PLC. Composition of tested mixtures was based on equal amount of sand and binder, only water was dosed in different amount [4]. Mixing of mixture was made in free fall concrete mixer therefore it was needed to monitor sufficient shuffle of all components. Specimens made of different anhydrite mixtures (without fibres, with different kind of fibres) were tested at laboratory of CTU. Following tests were made with indurated specimens:

- compression strength of tested specimens
- splitting tensile strength of tested specimens
- flexural strength (4-point bend) including scan of deflection for data evaluation of fracture energy and equivalent tension
- processing of results and evaluation of tests

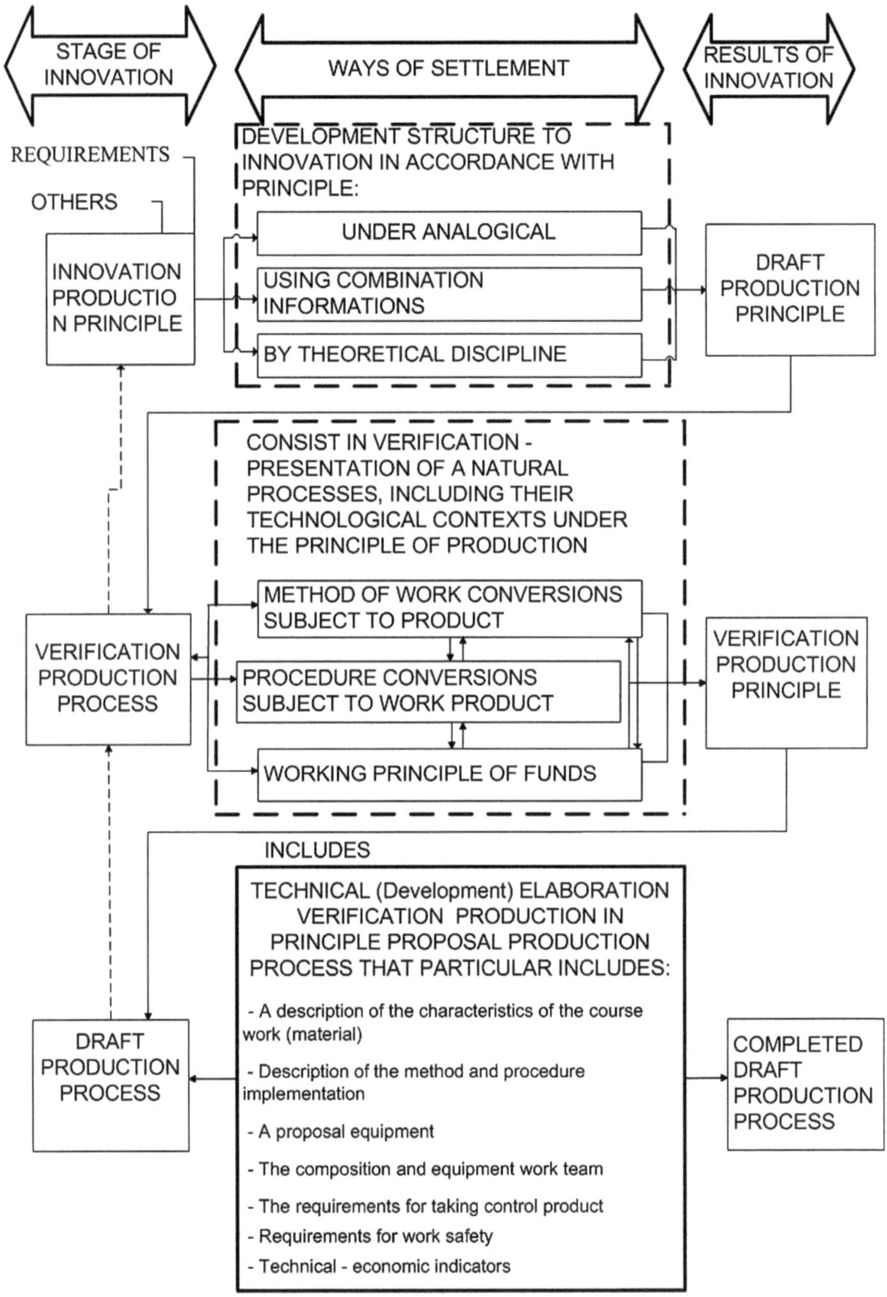

Fig. 2. Improvement of production process

Note: Results of those tests are at the end of the article.

Five mixtures of anhydrite screed (marked AP1-AP5 – see Table 1) were designed and experimentally tested. Referential mixture was selected mixture without fibres marked AP1. Other mixtures contain glass or polypropylene fibres. Following specimens were tested:

Table 1. Composition of tested mixtures

Composition of tested mixtures						
Dosed components	Units	AP1	AP2	AP3	AP4	AP5
Sand	kg	28	28	28	28	28
Binder	kg	11,1	11,4	11,4	11,4	11,4
Water	l	6,5	6,5	6,8	6,9	6,5
Glass fibres ANTI CRAK HP (length 12 mm)	g	–	20	–	–	–
Glass fibres ANTI CRAK HPL	g	–	–	250	–	–
Polypropylene fibres FIBRUCO (length 12 mm)	g	–	–	–	20	–
Polypropylene fibres ECONO MONO (length 19 mm)	g	–	–	–	–	20
Total amount	kg	45,9	45,92	46,45	46,32	45,92
Properties of the fresh mixture						
Spillage of cone for mortar tests	mm	250	240	220	200	210

- 15 × beam 40 × 40 × 160 mm
- 10 × beam 100 × 100 × 400 mm

Specimens were stored in laboratory conditions in temperature 20°C ± 5°C until start of the tests – i.e. 10.10.2005 (Fig. 3).

Final results of individual test are presented at the following tables. More detailed results are at the end of article (Table 3).

2.1 Specimens for Tests of Anhydrite Screeds: Conclusion

It is evident from finished laboratory tests of specimens, i.e.

- compression strength of tested specimens
- splitting tensile strength of tested specimens
- flexural strength (4-point bend) including scan of deflection for data evaluation of fracture energy and equivalent tension.

There are significant differences between specimens. Anhydrite screed specimens reinforced by fibres have better properties then specimens without fibres.

Table 2. Flexural strength test and compression strength test of the beam specimens 40 mm × 40 mm × 160 mm

Specimen's mark	Flexural		Compression strength			
	Force [kN]	Tension [MPa]	Force 1 [kN]	Force 2 [kN]	Tension 1 [MPa]	Tension 2 [MPa]
AP1-10	1,38	3,23	22,3	26,0	14,0	16,3
AP1-20	1,37	3,24	21,8	21,5	13,8	13,6
AP1-30	1,65	3,90	30,8	32,0	19,4	20,2
Ø		**3,46**				**16,2**
AP2-10	2,02	4,73	39,0	38,3	24,4	23,9
AP2-20	1,95	4,54	39,8	39,3	24,7	24,3
AP2-30	1,97	4,64	38,5	40,3	24,2	25,3
Ø		**4,64**				**24,5**
AP3-10	2,15	5,06	37,0	37,0	23,2	23,2
AP3-20	2,05	4,85	39,8	39,8	25,2	25,2
AP3-30	2,09	4,87	34,3	34,3	21,3	21,3
Ø		**4,93**				**23,2**
AP4-10	2,06	4,72	38,5	34,8	23,6	21,3
AP4-20	2,37	5,50	38,3	33,8	23,7	20,9
AP4-30	2,05	4,80	33,5	34,8	21,0	21,8
Ø		**5,01**				**22,1**
AP5-10	2,13	4,98	37,0	36,8	23,1	22,9
AP5-20	2,25	5,38	38,0	37,8	24,2	24,1
AP5-30	2,41	5,46	41,0	36,8	24,9	22,3
Ø		**5,27**				**23,6**

Fig. 3. Spillage test, series of specimens

Table 3. Four point test of the beam specimens 100 × 100x400 mm – assessment of fracture energy

Specimen's mark	D_n^C [Nmm]	$D_{fI,I}^f$ [Nmm]	$D_{fI,II}^f$ [Nmm]	Fractural energy G_f for different parts of graph [J/m²]				
				$G_{f,fu}$	G_f partial, I	G_f partial, II	$G_{f,I}$	$G_{f,II}$
AP1-1	904			90,4			90,4	90,4
AP1-2	904			90,4			90,4	90,4
Ø	**904**			**90,4**			**90,4**	**90,4**
AP2-1	1112	382	502	111,2	38,2	50,2	149,4	199,6
AP2-2	1103	375	473	110,3	37,5	47,3	147,8	195,1
Ø	**1103**	**378**	**488**	**110,3**	**37,8**	**48,8**	**148,1**	**196,9**
AP3-1	1346	1968	3748	134,6	196,8	375	331,4	706,2
AP3-2	1461	2206	4683	146,1	220,6	468,3	366,7	835
Ø	**1404**	**2087**	**4216**	**140,4**	**208,7**	**421,6**	**349,1**	**770,6**
AP4-1	1430	1659	1659	143	165,9	165,9	308,9	474,8
AP4-2	1167	522	781	116,7	52,2	78,1	168,9	247
Ø	**1299**	**1091**	**1220**	**129,9**	**109,1**	**122**	**238,9**	**360,9**
AP5-1	1530	1318	1318	153	131,8	131,8	284,8	416,6
AP5-2	1439	1356	1356	143,9	135,6	135,6	279,5	415,1
Ø	**1485**	**1337**	**1337**	**148,5**	**133,7**	**133,7**	**282,2**	**415,9**

3 Calculations of Critical Point of Proliferation of Breach

Since the cracks in anhydrite screed are the first reason of the board fracture, there were made other analyzes after evaluation of results (to compare possible influence of fibres with creation of cracks in the anhydrite screed):

- specimens without fibres and
- specimens with polypropylene fibres ECONO MONO – length of fibre 19 mm. Those specimens reach to the best properties (if we pass away fractural energy of specimens with ten times amount of glass fibres).

For computer simulation of anhydrite screed behavior was used model board with following properties:

- Width of the model board was 1 m, length 20 m, thickness 30 mm
- Board was put down on the elastic pad with max. compressibility of 5 mm with load of 2 kN.m²
- Fracture energy was taken over from tables above
- Stress modulus of anhydrite was used 30 GPa
- Computation was made in software Mathematica
- Depth of initial defect (see Fig. 4) – results from the computation of the depth of initial defect compared to fractural energy it has been set crack propagation at bending moment maximum value 226 Nm. Results from graph below:

- The lesser fractural energy is, the lesser is depth of scratch – the reason of initial defect
- Fractural energy of anhydrite screed without fibres is 90,4 J/m2 – depth of scratch is 24 mm on 30 mm thick board
- Fractural energy of anhydrite screed with polypropylene fibres length 19 mm is 282,2 J/m2 or 415,9 J/m2 – depth of scratch is 28 mm or 29 mm on 30 mm thick board

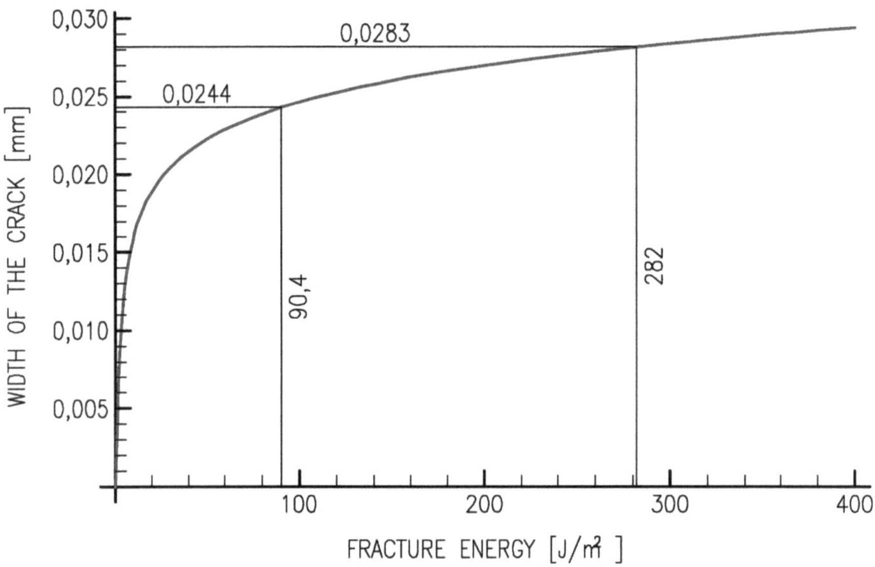

Fig. 4. Depth of initial defect (mm)

- Those ascertained results do not replay to practical knowledge, especially at anhydrite screed without fibres. In this case failure usually happen in the places where the scratch reaches 1/3 thickness of the board. During realization of anhydrite screed with polypropylene fibres of tested type there were no cracks in the rugged area of 48 m^2.

With regard to conclusions resulting from computer simulation based on fractural energy, comparison of scratch depth computation has been made.

In accordance to Table 2 there were determined average values of flexural strength of:

- anhydrite without fibres 3,46 MPa
- anhydrite with polypropylene fibres 5,27 MPa

from formula [1]:

$$\sigma = \frac{M}{W} \tag{1}$$

σ- flexural stress, W - section modulus, M - bending moment

$$W = \frac{1}{6}bh^2 = \frac{1}{6}1 \times 0,03^2 = \ 1,5 \times 10^{-4}\ \text{mm}^3$$

$$W = \frac{M}{\sigma}$$

$$\frac{1}{6}bh^2 = \frac{M}{\sigma} \quad \Rightarrow \quad h = \sqrt{\frac{6M}{\sigma \times b}}$$

(1) $\sigma = 3,46$ MPa

$$\Rightarrow \quad h = \sqrt{\frac{6 \times 226 \times 10^{-6}}{3,46 \times 1}} = 0,198\ \text{m} = 19,8\ \text{mm}$$

(2) $\sigma = 5,27$ MPa

$$\Rightarrow \quad h = \sqrt{\frac{6 \times 226 \times 10^{-6}}{5,27 \times 1}} = 0,1603\ \text{m} = 16\ \text{mm}$$

4 Conclusion

Resulting from comparison of used methods there are different criteria for crack propagation for method based on flexural strength and for linear fracture mechanics. According to [1] “Linear fractural mechanics is based on mathematical apparatus of flexible continuum and is trying to find quantitative characteristics. Those characteristics are decisive quantities for assessment of stability or lability of the crack”.

- According to computation of flexural stress it is possible to state – the crack begin to expand if stress achieves tensile strength (there is no concentration stress expected at the edge of the crack - pure bending)
- According to computation of fractural mechanics it is possible to state – the crack begin to expand when the depth of scratch achieves value of fractural energy (singular stress is infinite, it is not on the edge of the crack)
- The influence of fractural energy is minor for crack creation; major influence has flexural strength of material.
- Minimizing of crack creation should be reached by additions of fibres, i.e. increasing the tensile strength of material.

These conclusions are part of habilitation project of Ing. Miloslava Popenková, CSc. “Minimization of risk in the realization of anhydrite screed”.

References

1. Šafka, J., Hořejší, J.: Statické tabulky. SNTL, Praha (1987)
2. Kurth, H., Winter, K.H., Pospichal, V.: Ecological Requirements for Concrete Construction. Appl. Mech. Mater. **796**, 111–118 (2015)
3. Kurth, H., Winter, K.H., Pospichal, V.: Building condition record and cost estimation. Appl. Mech. Mater. **796**, 137–143 (2015)
4. Svoboda, P., Kurth, H.: Relation of Construction Works to Water Quality. Appl. Mech. Mater. **843**, 87–92 (2016)
5. Šejnoha, J., Kufner, V.: Pružnost, pevnost, plasticita III. ČVUT Praha (1990)

The Influence of Subsoil on Electromagnetic Detection

Ondřej Kašpar(✉)

Faculty of Science, Institute of Hydrogeology, Engineering Geology and Applied Geophysics, Charles University, 151 Engineer Battalion, Czech Army Forces, VÚ 6624 Bechyně, Czech Republic
kasparond@natur.cuni.cz

Abstract. Decommissioning of exercise ranges reflects downsizing of Czech Army. That process (started in 1993) requires disposal of all remnants of military activities, namely unexploded ordnance. Methodology of explosive ordnance clearance (EOC) is based on using of metal detectors.

The world wide experience witnesses some factors limiting or even eliminating their utilisation. Metal debris diminishes EOC reliability, when conducted in the outskirts of urban areas. Apart of this, there are cases when false alarms are observed without presence of any metal waste inside ground. This phenomenon was called "mineralisation of ground" initially. Term "mineralisation" was considered either as a content of mineralized groundwater or as a content of metallic minerals dispersed in rocks. However, the content of ferromagnetic minerals in rocks occurred under scientific interest. Research works have been conducted to clarify its nature and to recommend how to practise EOC procedures under those conditions.

Electromagnetic detection is based on the Faraday's rule of mutual induction between searching coil and a target. From this point of view the expression "mineralized soil" should be rectified. Even impregnation of ferromagnetic materials inside absolutely non-conductive rocks can cause false signal of detector. This is the right reason of "mineralisation". Where the bedrock consists of rocks containing magnetite, the success of electromagnetic searching occurs at risk.

Keywords: Electromagnetic detection · Mineralisation · Ferromagnetism · Bedrock

1 Introduction

Electromagnetic detection is based on the Faraday's rule of mutual induction. The searching coil of the detector approaches the target represented ammunition (preferably) which functions as a supplemental coil. Eddy currents can exist when size and shape of target allow it. The suitable expression for afterwards explanations should be derived from the following formula:

A. Kravcov et al. (eds.), *Durability of Critical Infrastructure, Monitoring and Testing*, Lecture Notes in Mechanical Engineering, DOI 10.1007/978-981-10-3247-9_11

$$\tilde{E} = j\omega L_1 I_1 \left(1 - k^2\right) \tag{1}$$

The product in front of the bracket of the expresses alternating electric field E inside of searching coil given by current I1, frequency ω, intrinsic induction L1. The symbol j (imaginary unit) indicates out of phase status between E and I1. The expression in the brackets indicates effect of mutual induction (interaction) between target and searching coil. When no interaction occurs, k → 0 and intrinsic induction of searching coil remains. When interaction occurs, k → 1 and mutual induction affects the E inside searching coil in such way that E is eliminated subsequently. This expression is correct when conductivity affects interaction between coils entirely. Ferromagnetic materials (i.e. steel) produce a interaction based on the magnetic permeability.

2 Principles of Electromagnetic Detection

Basic equations of electromagnetic field are following:

$$rot\vec{H} = \vec{j} + \frac{\partial}{\partial t}\vec{D} \quad Amper's\ rule \tag{2}$$

$$rot\vec{E} = \frac{\partial}{\partial t}B \qquad Faraday's\ rule \tag{3}$$

where:

$$\vec{j} = \sigma\vec{E} \quad \sigma : conductivity$$
$$\vec{D} = \varepsilon\vec{E} \quad \varepsilon : permitivity$$
$$\vec{B} = \mu\vec{H} \quad \mu : permeability$$

E, H vectors of electric, magnetic intensity
D, B vectors of electric, magnetic induction.

3 Phenomenon "Anomaly" in Electromagnetic Field

Electromagnetic field of sources existing inside absolutely non-conductive media is called primary field. The rock massif on resistivity over 1000 Ωm shall be considered non-conductive as well. Calculations of primary field are basically simple. The deviations from the primary field are anomalies. Anomaly is caused by local conductive formations, generating eddy currents. The surrounding non-conductive media the magnetic component prevails thus eddy currents shall be generated by magnetic field. Response parameter is $\alpha = \frac{\omega L}{R} \approx \sqrt{\omega\mu\sigma} \cdot r$ where r shall be separation between coplanar coils. Consequently, anomaly of magnetic type is formulated. The normal field is close to primary field, if meets the condition:

$$\alpha = \sqrt{\omega\mu\sigma} \cdot r \prec 1 \qquad (4)$$
$$\alpha : \textit{response parameter}$$

This condition is called proximity zone. The assignation of the frequency ω and distance r feasible to meet the condition can keep the normal field close to primary field. In the case of anomaly the condition is misbalanced and the anomaly is indicated. Eddy currents are concentrated in local conductive formations causing anomalies of electric type. They are difficult to be quantified bus relatively simple to be registered.

4 Propagation Constant γ

Propagation constant γ characterizes an effect of the homogenous media of determined parameters. The densities of the conduction currents and displacement currents are defined by the second row of the equations.

$$\gamma = \sqrt{\omega^2\mu\varepsilon + i\,\omega\mu\sigma} \quad \text{Propagation constant } \gamma \text{ characterizes an effect of the homogenous media} \qquad (5)$$

$$\sum j = j + j_D = \sigma E + \frac{\partial}{\partial t} D = \sigma E + i\omega\varepsilon E \quad \text{Density of conduction currents and displacement currents} \qquad (6)$$

$$\frac{\mathrm{Im}\, k^2}{\mathrm{Re}\, k^2} = \frac{\omega\mu\sigma}{\omega^2\mu\varepsilon} = \frac{\sigma}{\omega\varepsilon} = \frac{j}{j_D} \quad \text{Ratio of second powers of the propagation constant components} \qquad (7)$$

$$If : \frac{\sigma}{\omega\varepsilon} \prec\prec 1 \Rightarrow \gamma = \omega\sqrt{\mu\varepsilon} \quad \text{Propagation constant in the case of displacement currents predominance} \qquad (8)$$

$$If : \frac{\sigma}{\omega\varepsilon} \succ\succ 1 \Rightarrow \gamma = \sqrt{i\omega\mu\sigma} \quad \text{Propagation constant in the case of conduction currents predominance} \qquad (9)$$

The third row defines a ratio of second powers of the propagation constant components and defines them as fraction of the conduction and displacement currents. The last two rows define a propagation constant in the case of conduction currents predominance (above) and vice versa (below). Predominance of conductive currents defines (quasi) conducting region, where displacement currents could be ignored. Predominance of displacement currents defines (quasi) dielectric region.

Real part of the propagation constant is associated with phase-shifting while imaginary part is associated with the attenuation coefficient.

Propagation constant for conductive and non-conductive media were considered in connection with representative soils property data. The average geological conditions manifest conductivity higher than 10–5 Sm^{-1} (corresponding 105 Ωm). The limit of

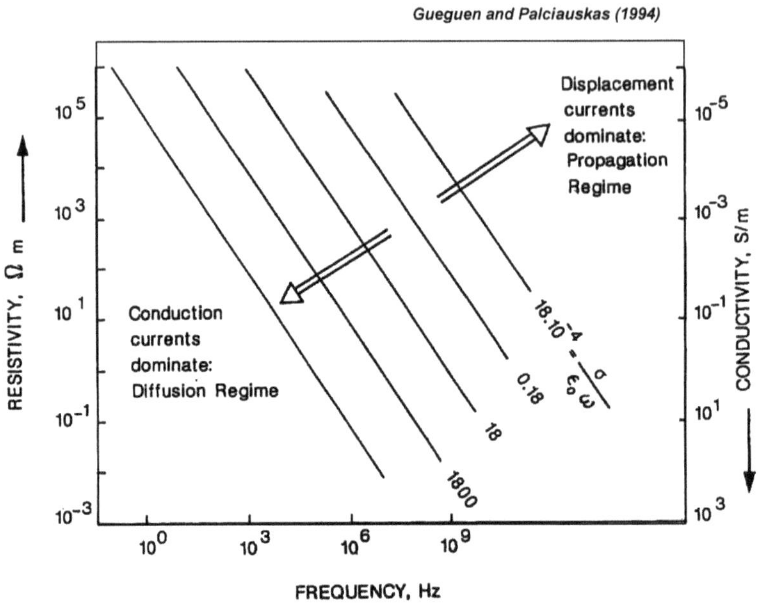

Fig. 1. Non-attenuated wave propagation for non-conductive media and attenuated mode for conductive media. By [2]

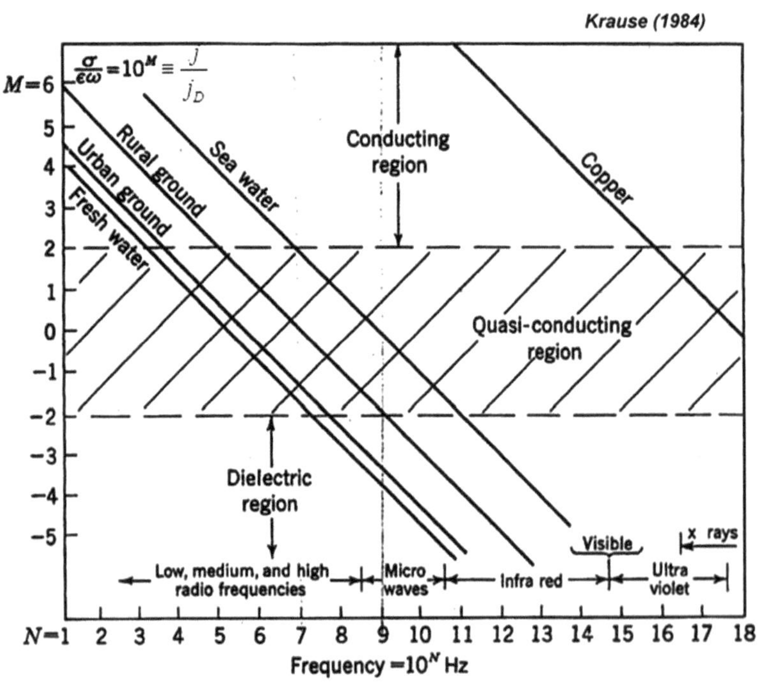

Fig. 2. Conducting, quasi-conducting and dielectric regains for certain frequencies and materials. By [2]

the Fig. 1 should be put on the ratio of 0, 18 between conductive and displacement currents.

Note, resistivity of media about 1000 Ωm allow quasi conductive approximation for frequencies reaching almost 1 GHz. As suggested by Fig. 2, modelling of the microwave instruments response with representative soil parameters demonstrated that even at radar frequencies appropriate for mine detection (1 GHz), the dielectric approximation is not generally obligatory.

5 Ferromagnetism

In connection of metal detectors the (quasi) conducting approximation has been accepted, where displacement currents are negligible.

Ferromagnetic materials are characterized by the complicated course of magnetization/demagnetization. This whole cycle is characterized by hysteresis loop representing energy spent for reorientation of the magnetic domains (Table 1).

Table 1. Explanation of vector components of magnetisation. By [7]

Magnetic field in absence of the ferromagnetic materials (i.e. air)	$B = \mu_0 H \quad \mu_0 = 4\pi 10^{-7}\ \mathrm{Hm}^{-1}$	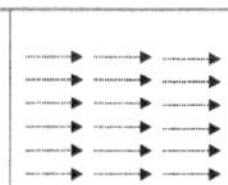
Magnetic field in presence of ferromagnetic materials possessing induced magnetisation only	$B = \mu_0(H + M_i) = \mu_0(H + \kappa H) = \mu_0 H(1 + \kappa) \equiv \mu_0 \mu_r H$	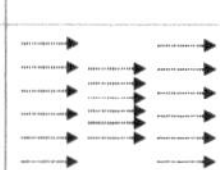
Magnetic field in presence ferromagnetic materials possessing both induced and permanent magnetisation	$B = \mu_0(H + M_i + M_r) = \mu_0(H + \kappa H + M_r)$	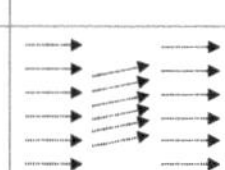

Relations, when $\mu = \mu_0$: When $\alpha \to 0$, ($\omega \to 0$, $\sigma \to 0$, note $R \to \infty$), the secondary field is referred to the resistive limit, secondary field is phase shifted by $-\pi/2$ relative to primary field.

When $\alpha \to \infty$, the secondary field is referred to the induction limit. Eddy currents are restricted to the surface of conductive medium (skin effect) and the secondary field cancels the primary field.

Relations, when $\mu = \mu_0\ \mu_r$: When $\alpha \to 0$, response parameter dominated by magnetisation, secondary field is phase shifted by 0. Increasing frequency and conductivity, the polarity becomes reversal of in-phase component consistent with

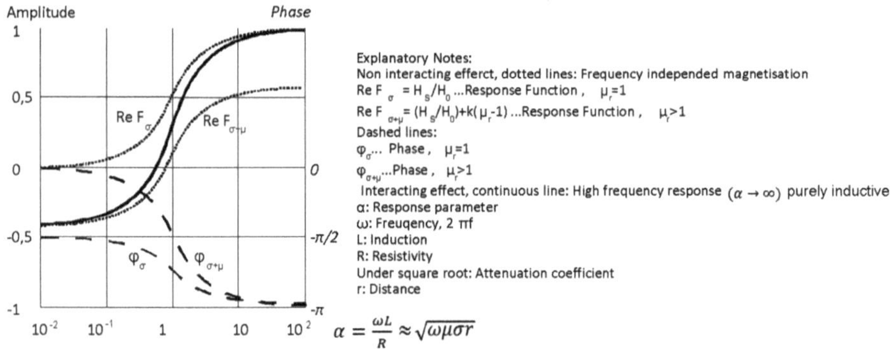

Fig. 3. In-phase inductive response function F dependent on the response parameter α for conductive-magnetic ground. By [1]

transition from magnetisation to induction dominated response. The induction response is enhanced by magnetisation of the soil.

Effect of eddy currents on ferromagnetic material indicates next figure. Alternating magnetic flow Φ_1 induces eddy currents acting against Φ_1 by their magnetic field Φ_2. Resulting magnetic flow is diminished. This effect exhibits itself like reduction of the permeability of the ferromagnetic material (Fig. 4).

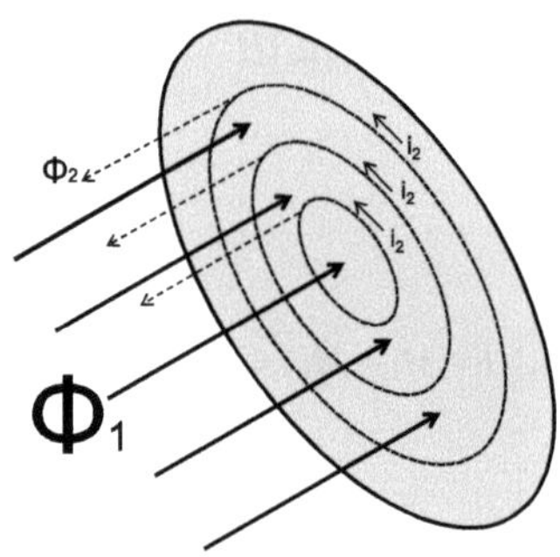

Fig. 4. Magnetic effects of eddy currents. By [17]

The final expression describing interaction between searching coil and the target of ferromagnetic material could be expressed:

$$E_h = \sum \left(E_{MAG} + \tilde{E}\right) \quad where: \quad E_{MAG} = rotM_i \\ where: M_i \ldots Induced\ magnetisation\ vector \tag{10}$$

Limiting high frequency response (α → ∞) for conductive/magnetic ground remains purely inductive, consistent with –π shift. The secondary magnetic field associated with saturation eddy currents cancels the primary field within the medium (permeable soil) and **consequently nullifies related influence of magnetisation.**

From this expression could be derived those results:

- Electromagnetic detector can search for targets of ferromagnetic and non-ferromagnetic materials.
- Presence of eddy currents limits searching for targets of non-magnetic materials.
- Shape of non-magnetic materials eliminating presence of eddy currents causes those materials undetectable.
- Ferromagnetic materials could be detected notwithstanding of eddy currents presence.
- Even impregnation of ferromagnetic materials inside absolutely non-conductive rocks or soils can cause signal of detector.

6 Natural Limitations of Metal Detection

Some manuals (i.e. [3]) refer to relation of red colouring of the topsoil and adverse conditions for detection. They justify it by the presence of mineralized soil containing ferromagnetic minerals. This explanation cannot be acceptable without reservations. The red colouring inside of strata complex indicates arid climate, where evaporation prevail precipitation process. Simplified explanation can state, that capillarity force moved leached Fe^{2+} ions from parent rock upwards, where they oxidize. Accessory amount of haematite causes this colour. But this process is not attributable to magnetic parent rock exclusively. However, even nonmagnetic rocks like granite, gneiss or limestone contain minerals with Fe in their molecules. Recently, this process of oxidization is visible in Karst, where defiles constructed for new roads are excavated through calcareous rocks.

Almost opposite process could be observable in moderate humid climate ruling in the mid Europe. The precipitation prevail evaporation here while leached Fe ions are moved by rain water downwards, to the subsoil layer called illuvium. No reports notice it as a “mineralized soil” despite of its real mineralisation in the comparison of other pedogenic layers.

7 Ground Conductivity

The skeleton of soil acts as insulator. The electric conductivity is caused by minerals leached from the soil and dissolved in groundwater. The changes of ground conductivity are closely connected with permeability of soil. The arenaceous materials (sand, gravel) are pervious, thus groundwater cannot mineralize itself and its conductivity is low. On the contrary, argillaceous materials (clay, marl) are impervious; able to keep groundwater among grains, thus mineralisation of groundwater and its conductivity is high, comparing the previous instance.

Operation manuals (i.e. [4]) mention an effect of highly saline sea water, namely at coastlines, requiring special provisions for detection. With respect to Figs. 2 and 3 this recommendation looks controversial. Even in the armament of Czech Engineers the Aquanaut detector is listed, designed for sea divers. The Authors suppose that the recommended provisions should be valid for "black beaches" where the sand is a weathering product of volcanic rocks (see hereinafter).

The previous experience of the Authors indicates that grain size and humidity of soil is easily to be balanced for detectors.

8 Ground Magnetisation

A series of ferromagnetic minerals induce magnetic properties of soils. Basically, magnetite is the main contributor to magnetize soil. If present in sufficient amount, leads to magnetic susceptibilities high enough to have a negative effect on the performance of metal detectors [11]. High initial content of magnetite is often associated with basic magmatic rocks.

Basalt is well known representative of those rocks. For the purpose of this technical information the term basalt will substitute all magnetite containing rocks, notwithstanding of position, origin or metamorphism inside of rock massif. For operators shall be sufficient to define "basalt type rock". Its regolith contains magnetite as a relic of the parent rock (Fig. 5).

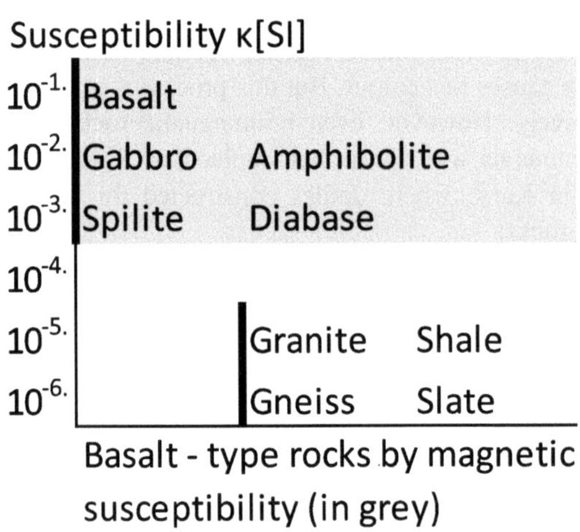

Fig. 5. Comparison of susceptibility between basalt type rocks and the other ones. Assembled by Kašpar.

Note: Referring to intermediate conclusions from [11], the level of the leaching process causing magnetite decay should be omitted due to its unpredictability.

9 Attenuation Rate

Attenuation of the electromagnetic field inside rock massif could be demonstrated by conductive insertion of the alternating electric current between two electrodes connected by cable.

Figure 6 shows situation when AC is grounded by electrodes AB. Cable connecting AB conducts J0 alternating current and ground is affected by alternating current J, distributed by the depth of electric flow. Magnetic field H is generated around the cable inducting current Jind.

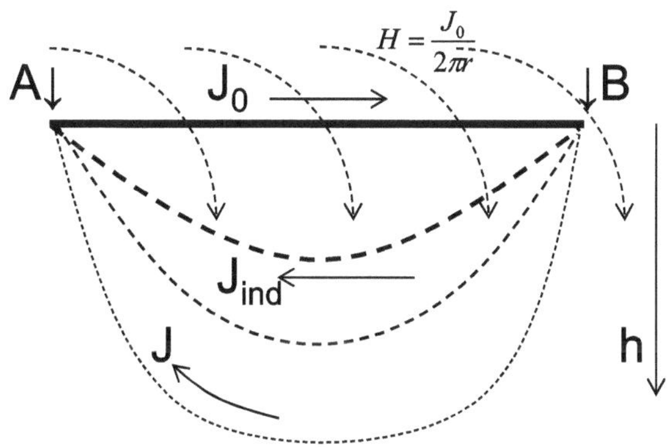

Fig. 6. Distribution of conductive/induction currents inside ground depending on distance AB. By [9] simplified by Author.

Intensity of inductive electric field depending on the resistivity of ground presents following Fig. 7

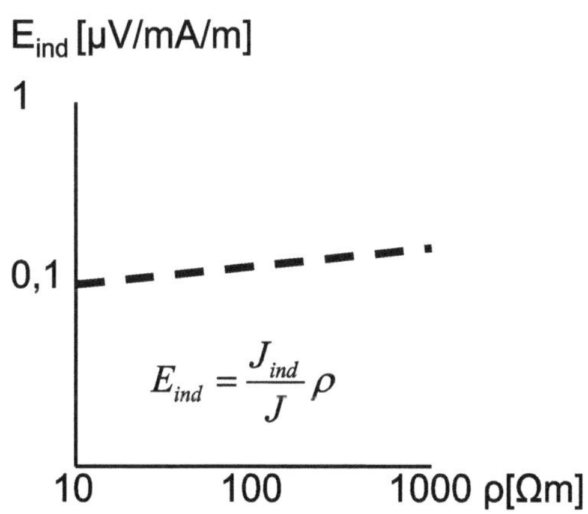

Fig. 7. Relation of the E_{ind} on the ground resistivity. Assembled by Kašpar.

There is a significant fact that its dependency on electric resistivity of ground should be almost neglected. The resistivity is calculated as a voltage to current ratio. The voltage shall be observed between two potential electrodes (dipole) positioned in the vicinity of electrodes AB. The mutual position of both electrode pairs characterizes a coefficient z, as seen from following expression. When the potential dipole is affected by inductive component mainly (AB distance very long), then electric resistivity observed cannot characterize the correct situation of the ground.

$$\rho_\omega = z\frac{\sqrt{U_{ind}^2 + U^2}}{J} \tag{11}$$

$$U \prec\prec U_{ind} \Rightarrow \rho_\omega = z\frac{U_{ind}}{J} = z\frac{J_{ind}}{J}\rho = zE_{ind} \tag{12}$$

$$E_{ind} \neq f(\rho) \Rightarrow \rho_\omega = f(z) \approx f(h) \tag{13}$$

Finally, the electric resistivity becomes a product of coefficient z and the intensity of inductive electric field E_{ind}. The previous figure shows that intensity does not depend on the resistivity, and then the resistivity of ground depends on the coefficient z, comprising geometry of the measurement and thus the depth h. This relation can be presented as a line dipping by 63,5° on the bi logarithmic scale, so called induction asymptote. This has been published by Vishnjakov (1966), who derived it for the frequency 20 Hz. Next text can explain the process how to make use of this induction asymptote to derive them for any frequencies.

$$E_h = E_0 e^{-\gamma h} \Rightarrow \frac{E_h}{E_0} = \frac{1}{e} \Rightarrow h = \frac{1}{\gamma} \tag{14}$$

$$\gamma = \gamma'' = \sqrt{\frac{\omega\mu\sigma}{2}} \tag{15}$$

$$h = \frac{1}{\sqrt{\frac{\omega\mu\sigma}{2}}} = \frac{1}{\sqrt{\frac{2\pi f 4\pi 10^{-7}\sigma}{2}}} = 500\sqrt{\frac{\rho}{f}} \tag{16}$$

The depth penetration is a route through a massif, where the electromagnetic field is diminished e-times. The inductive asymptote and relation (h, σ, f) can establish an induction asymptote for any frequency based on proportionality principle:

$$\sqrt{\frac{\rho}{f}}(20\text{ Hz}, 10\ \Omega\text{m})\ldots 100\text{ m}(Vishnjakov) \tag{17}$$

$$\sqrt{\frac{\rho}{f}}(X\text{ Hz}, 10\ \Omega\text{m})\ldots\ldots proportionality \tag{18}$$

Electromagnetic wave impacting conductive surface is attenuated by relation e-γh. γ is an imaginary component of the propagation constant.

Induction asymptote presented on Fig. 8 is followed by appropriate frequency. 20 Hz asymptote has been confirmed by Vishnjakov, 700 Hz by Author. Those confirmations were achieved by symmetric vertical electric sounding (VES) and are positioned in highlighted rectangle (grey). Radar frequencies have been established for 0,1, 1, 10 and 100 GHz.

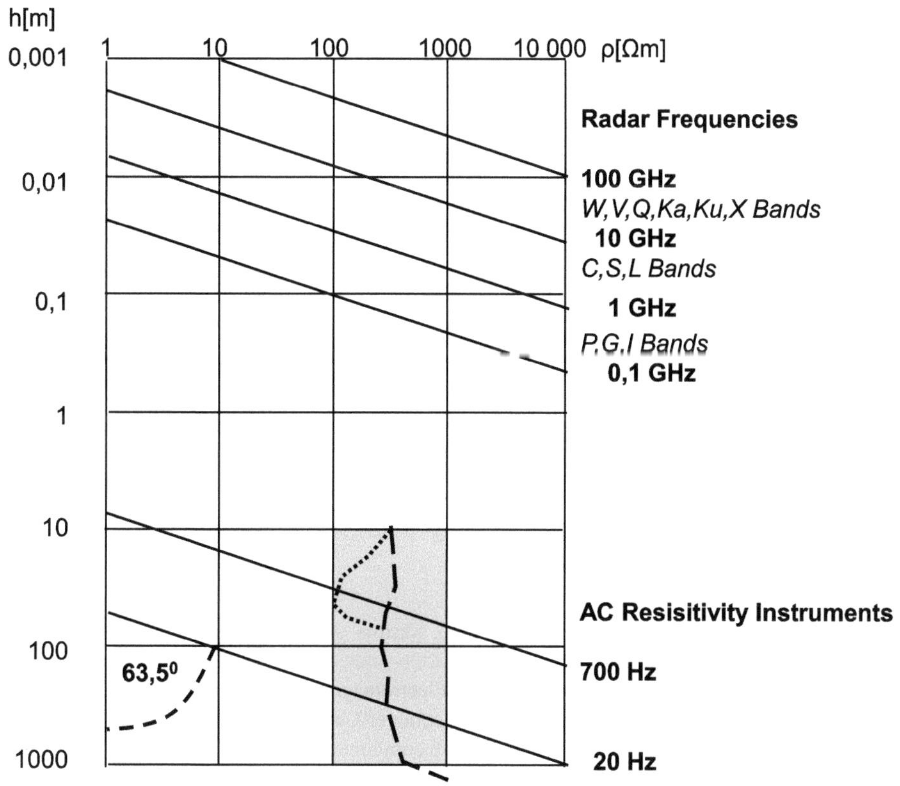

Fig. 8. Induction asymptotes corresponding to certain frequencies. By [7] simplified by Author.

The complete difference between electric sounding (based on electric resistivity) and radar prospection (based on permittivity) can produce doubts on correctness of this application of proportionality principle. With reference to Fig. 8 the statement based on average resistivity (urban/rural ground) and quasi conducting region corresponding with frequency 1 GHz as correct can be presented.

Comparing induction asymptote and formula h the last one provides results approx. 5 times deeper. It is difficult to explain this difference but with regard to VES results, the induction asymptote shall be considered as a limit, from it the skin effect affects measured results.

10 Conclusion

Metal detectors are indispensable tools to achieve the decommissioning works, demining or disposal of explosive ordnance successfully.

Ferromagnetism of rocks affects ammunition detection. The presence of plutonic rocks, like gabbro or peridotite, volcanic rocks like basalts, metamorphic rocks like greenstones or amphibolite downgrades efficiency of metal detectors. The Author are aware of potential misunderstanding, when petrology nomenclature could confuse operators or planners. To simplify this problem, this article defined non generic term "basalt type rock". This term can overarch all kind of rocks containing magnetite regardless of their origin. Some statements incorporated in operational manuals, indicate an effect of so called "mineralisation". This article rectified these statements. Electric conductivity was mentioned and red colouring of soil has been explained.

Recently, the remarkable progress of research works has been done in theory and prognosis as well. Research results (see [2, 11]) turned interest to bedrock and soil petrology. This article transferred those research results to Bohemian Massif structures and exercise ranges locations. The coincidence of "basalt type rock" and location of exercise ranges inside Czech territory has been studied.

The coincidence of potential EOC works and location of basalt type rock shall be the subsequent necessary step. Measurements of magnetic susceptibility on outcrops of above mentioned "basalt type rocks" would confirm or reject prediction of detectors deficiency. List of maps of bedrock is strongly recommended, preceding all planning documents related to decommissioning process.

References

1. Beyr, P.: Ammunition Detection (in Czech): Skripta. First, VA Brno, Brno (2003). ISBN 11/15 PSP 568
2. Cross, G.: Terrascan geophysics. In: Soil Electromagnetic Properties and Metal Detector Performance: Theory and Measurement, Vancouver Canada, November 2008
3. Ground mineralisation. http://www.minelab.com/emea/consumer/knowledge-base/terminology. Accessed 15 May 2016
4. How does DEUS deal with ground mineralisation and how should I adjust the ground effects compensation in the GROUND menu, in manual, pumping or tracking modes? Is it necessary to adjust? FORUM XP (2009). http://www.xpmetaldetectors.com/xpforum/viewtopic.php?f=14&t=27. 15 May 2016
5. How does ground mineralisation affect metal detecting? FORUM XP (2009). http://metaldetectorsa.co.za/. Accessed 18 June 2016
6. Hrouda, F., Chlupáčová, M.: A martin chadima. Advanced geoscience instruments. In: The Use of Magnetic Susceptibility of Rocks: Case Histories Study, Brno, Czech Republic (2009)
7. Karous, M., Kněz, J.: Geoelectric methods of prospection. In: Electromagnetic Methods (In Czech). SPN, Praha (1977). 280 s

8. Maňas, P., Mazal, J.: The reach-back concept in the czech army corps of engineers. In: International Conference on Military Technologies. Faculty of Military Technology, University of Defence, Brno (2009). ISBN 978-80-7231-648-9
9. Mareš, S., et al.: Applied Geophysics, Introduction (in Czech). First, 591 p. SNTL, Praha (1979).
10. Mareš, S., et al.: Geophysical Methods in Geology and Engineering Geology (in Czech), no. 1. SNTL, Praha (1983). 200 s
11. Mcdermott, M.: Highly mineralized soil and ground balancing. In: Hollands Brook (2011). http://www.hollandsbrook.com/blog/mineralized-soil-ground-balance/ Accessed 18 July 2016
12. Preetz, H., Altfelder, S., Hennings, V., Igel, J., Harmon, R.S., Thomas Broach, J., Holloway, J.H., Jr.: Detection and sensing of mines, explosive objects, and obscured targets, case study of Angola XIV. s. 730313-730313-11 (2009). doi:10.1117/12.819394. http://proceedings.spiedigitallibrary.org/proceeding.aspx?articleid=1338053
13. Příhoda, K.: Modern Mine Detectors. Principles and Real Capabilities (in Czech). ATM (1996). č. 196, s. 3
14. Schuler, W.: Prospecting Instruments for Professionals (in Czech). BEN, První české vydání. Praha (2000). 136 s. ISBN 80-86056-85-6
15. Nato. Stanag 4587: Close in Landmine Detector Test Procedures, 1 edn., 108 p, [s.l.]: [s.n.], 25 April 2007
16. United Nations Mine Action Centre: Report summary: findings and recommendations of the report on the hand held metal (mine) detector trials conducted in Bosnia during January 1997. Marshal Tito Barracks BFPO 543 England, Sarajevo, Bosnia, March 1997
17. Batka, V.: Elektromagnetismus (in Czech), no. 1. Práce, vydavatelství a nakladatelství ROH, Praha (1970). Polytechnická knižnice. ISBN 24-040-70

On Calculation of the Bearing Capacity of Self-opening Ground Anchors Using PLAXIS 2D Software Package

Ivan E. Sas(✉), Dmitry V. Morozov, and Nikolay A. Morozov

The National University of Science and Technology MISiS,
Leninskiy Prospekt 4, 119991 Moscow, Russian Federation
ivan_sas@bk.ru, moroz_off_123@mail.ru,
morozov499@yandex.ru

Abstract. The bearing capacity of a self-opening ground anchor is calculated using the empirical method and the plaxis software package. Stresses and strains can be determined in each element of the anchor using the approach based on the use of modern software packages enables to evaluate the stress-strain behavior of soil-structure system (geotechnical conditions taken into account), as well as the behavior of soil-anchor system as a whole and its individual elements. It is possible to estimate stresses in the area affected by the ground anchor, as well as the stability of the soil-anchor system as a whole.

Keywords: Anchor systems · Bearing capacity of anchor · Numerical simulation · Estimation of anchoring forces

1 Introduction

Construction of modern buildings and structures is often associated with a number of difficulties. First of all, these difficulties are related to ever-growing complexity in the structural design of buildings and structures and the complex geotechnical conditions in the areas allocated for construction projects. In this connection, geotechnical engineering is becoming an integral part of construction activities with a view to ensuring the stability of buildings and structures and the surrounding environment.

Among geotechnical engineering tools are anchor systems, including grouted anchors, prestressed anchors, screw-in anchors, etc. The interaction between anchors, building envelope, and surrounding soil results in the non-uniform stress-strain behavior of both the soil and the anchors in space and time. Therefore, the stress-strain behavior of soil-anchor-envelope system should be quantitatively assessed in order not only to determine the bearing capacity of the anchors and time-dependent changes in the tensile force in the anchor rods, but also to evaluate the effectiveness of geotechnical protective measures.

Nowadays ground anchoring systems are becoming increasingly popular. These systems have several advantages: there is no necessity to construct an anchor slab and drill holes; such anchors can be quickly and easily installed; there are various types of anchors suitable for different design solutions. In addition, anchors of this type are cost-effective.

A. Kravcov et al. (eds.), *Durability of Critical Infrastructure, Monitoring and Testing*,
Lecture Notes in Mechanical Engineering, DOI 10.1007/978-981-10-3247-9_12

However, calculation of the bearing capacity of such anchors is still a non-trivial task. On the one hand, there are several empirical approaches to evaluate the bearing capacity of anchors. But neither the complexity of geological conditions nor the fact that anchors of this type are of multi-element design are taken into account at that.

On the other hand, modern software packages (such as PLAXIS, ZSOIL, etc.) with their specially designed tools make it possible to evaluate the stress-train behavior of soil-structure systems, taking anchoring systems into consideration. Such an approach allows the complexity of geological environment to be taken into account. However, these tools exclude the possibility of taking the multi-element design of anchor systems into account.

The choice of a calculation method is dependent on the level of responsibility of buildings and construction site conditions. However, engineering design software packages make it possible, through the use of special built-in tools, to regard anchors of this type as multi-element objects i.e., design and simulate different parts of the anchors using different software tools.

2 Design Features of Self-Opening Ground Anchors

Often made of rolled metal, ground anchors are multi-element devices consisting of the following (Fig. 1): an anchor proper (self-opening element), an anchor bar/cable, and a locking stick/nut.

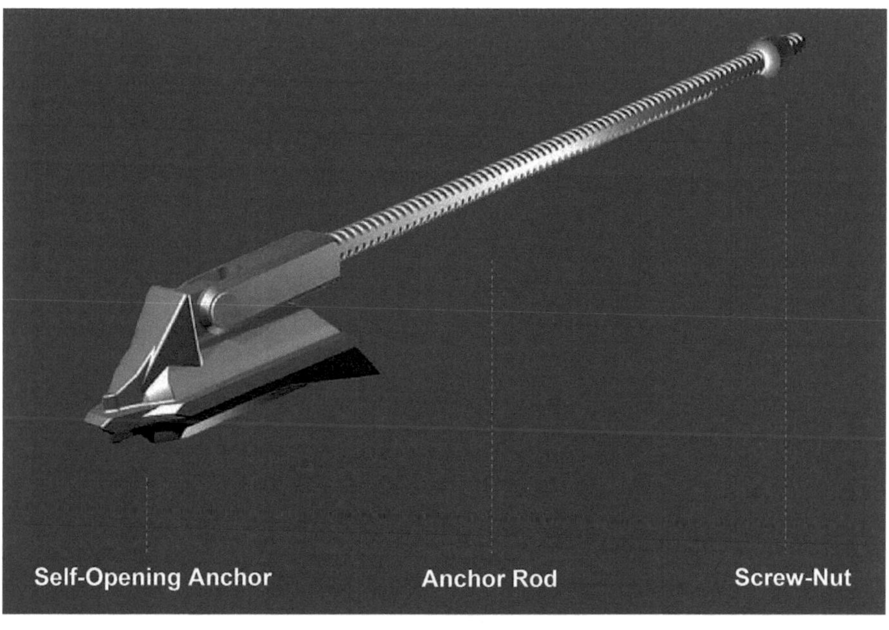

Fig. 1 Scheme of self-opening anchor

This design suggests that there are several “working” elements each characterized by its own strength properties and ultimate load capacity. Therefore, it is unreasonable to regard such an anchor as a single element when calculating its bearing capacity.

3 Classical Approach to Calculation of the Bearing Capacity of Self-opening Ground Anchors

The classical approach to calculation of the bearing capacity of self-opening ground anchors involves a theoretical solution to the problem of pulling out an anchor having a certain area and placed at a certain depth. Under the statement of the problem, when a ground anchor is subjected to a pull-out force, its stability is determined by the resistance (adhesion) of the surrounding soil mass affected by the anchor under a load applied at angle φ (angle of internal friction) (Fig. 17). In 2D, the bearing capacity of a ground anchor is estimated as follows:

- maximum magnitude of forces acting on the ground anchor is determined;
- the depth of the anchor is estimated;
- the area of ground surface (truncated cone) affected by the ground anchor is calculated;
- the shear resistance is evaluated from the calculated surface area on the basis of the physical-mechanical properties of the soils;

It is thus possible to evaluate the bearing capacity of the anchor from the shear resistance of the soil mass affected by the anchor. However, this scheme does not allow

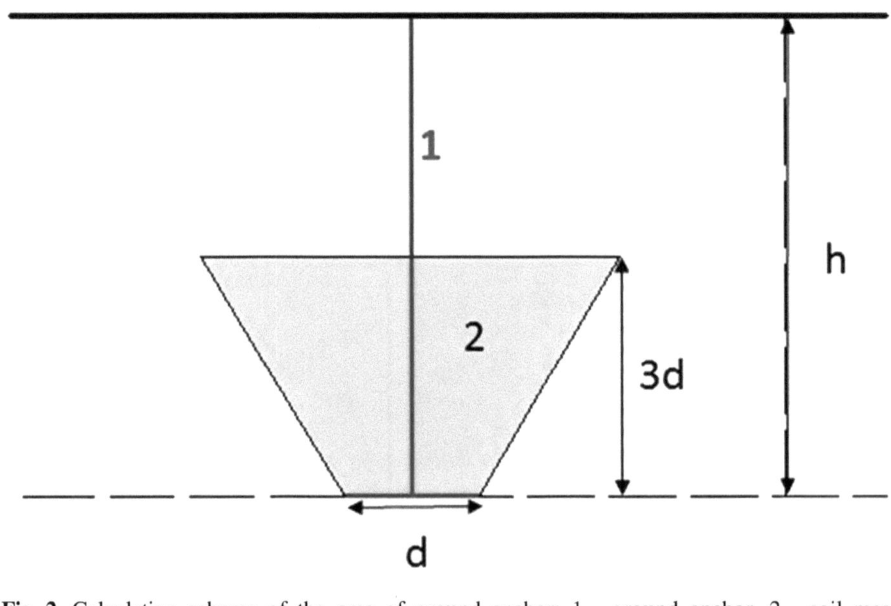

Fig. 2 Calculation scheme of the area of ground anchor: 1 - ground anchor, 2 - soil mass affected by anchor under pull-out load

the complexity of geology to be taken into account; it is clear from the scheme that the behavior of the anchor bar cannot be evaluated, which can be very important in some cases (Fig. 2).

4 Technique for Evaluating the Bearing Capacity of Ground Anchors Using PLAXIS 2D Software Package

As previously mentioned, modern software packages allow us to regard ground anchors as multi-element devices, i.e. take their design into account.

Given below is the calculation performed using PLAXIS 2D software package. A ground anchor was simulated by a two-element object consisting of:

- an anchor proper, i.e. a self-opening element that is firmly fixed in the ground. Since this element is rigidly fixed, it was simulated using the *fixed-end-anchor* tool *(with dimensions as per the anchor specifications).* This tool is a point and non-deformable one, which corresponds to the behavior of the anchor in operation.
- a steel bar, an element that takes a tension force and transmits it to the anchor. This element was simulated using the *steel bolt anchor* tool; this is fully consistent with the behavior of a steel bar under load.

It is clear from Fig. 3 that this approach is quite justified because the geotechnical conditions are taken into account when analyzing the behavior of the ground anchor. In addition, the stress distribution indicates that the ground anchor concentrates stresses which the steel bar transmits to it. Thus, the two-element model is more accurately reflects the behavior of a real ground anchor than the design model used in the empirical approach.

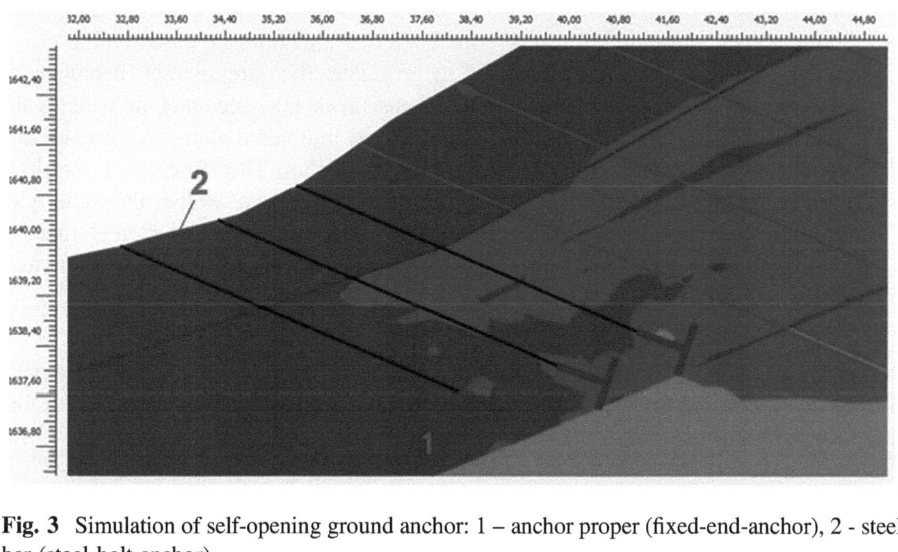

Fig. 3 Simulation of self-opening ground anchor: 1 – anchor proper (fixed-end-anchor), 2 - steel bar (steel bolt anchor)

It should be noted that the coupling between the *fixed-end-anchor* and *steel bolt anchor* elements is rigid, which corresponds to the actual design.

All the above considered, it may be stated that this approach makes it possible to estimate the stress-strain behavior of the soil-structure system, as well as the behavior of the soil-anchor system as a whole and its individual elements. Using the PLAXIS software package, stresses and strains in each element of the anchor can be evaluated. Therefore, it is possible to estimate the anchoring forces, as well as the stability of the soil-anchor system as a whole. In addition, data on the state of each element of the anchor can be used in non-destructive testing of anchor systems as a normal characteristic of their behavior.

5 Conclusions

The bearing capacity of a self-opening ground anchor was calculated using the empirical method and the PLAXIS software package; the following conclusions are made:

- Calculation of the bearing capacity of ground anchors is not an easy task: on the one hand, there are a number of empirical procedures to estimate their bearing capacity but these procedures ignore the complexity of geological conditions. On the other hand, modern software packages (such as PLAXIS, ZSOIL, etc.) with their specially designed tools make it possible to estimate the stress-strain behavior of soil-structure systems with regard to anchoring systems.
- Using the empirical method, the bearing capacity of a ground anchor can be determined from the shear resistance of the soil mass affected by the anchor. However, this scheme ignores the complexity of geology and, as it follows from the calculation scheme, the behavior of the anchor bar, which can be very important in some cases.
- All the above considered, it can be concluded that the approach based on the use of modern software packages enables to evaluate the stress-strain behavior of soil-structure system (geotechnical conditions taken into account), as well as the behavior of soil-anchor system as a whole and its individual elements. Stresses and strains can be determined in each element of the anchor. Therefore, it is possible to estimate stresses in the area affected by the ground anchor, as well as the stability of the soil-anchor system as a whole. In addition, data on the state of each element of the ground anchor can be used in non-destructive testing of anchor systems as a normal characteristic of their behavior.

Acknowledgements. The work was carried out with financial support in a part from the Ministry of Education and Science of the Russian Federation in the framework of Increase Competitiveness Program of NUST «MISiS» (№ K1-2015-025).

References

1. American Society of Civil Engineers: Subsurface investigation for design and construction of foundations of buildings, part I. J. Soil Mech. Found. Div. ASCE, **98**(SM5), 197, 481–490
2. Handy, R.L.: Geotechnical Engineering: Soil and Foundation Principles and Practice, 5th edn. The McGraw-Hill Companies, New York (2007). ISBN 9780071481205
3. Hoek, E.: Practical Rock Engineering/The Rocscience Website. Hoek's Corner (2006)
4. Meshchaninov, S.K.: Methods for Modeling and Managing Reliable Operation of Mining. National Mining University, Donetsk (2012)
5. Kehew, A.: Geology for Engineers and Environmental Scientists, 3rd edn. Pearson New International (2013). ISBN: 9781292039107
6. Plaxis 2D AE. User guide (2014)
7. Shapiro, D.M.: Theory and Computational Models of Bases and Facilities Geotechnics. Publishing and Printing Center «Scientific book», Voronezh (2012)

Experimental Ballistic Loading of Steel Fiber Reinforced Concrete Slabs and Unreinforced Concrete Slabs by Plastic Explosives

Jiří Štoller and Petr Dvořák(✉)

University of Defence, Kounicova 65, 662 10 Brno, Czech Republic
{jiri.stoller, petr.dvorak}@unob.cz

Abstract. The article presents the results of experimental tests carried out on steel fiber reinforced concrete slabs and on unreinforced concrete slabs. The aim of the experimental tests was to determine its suitability for anti-spalling layer by using effects of plastic explosive PlHx 30 explosion. Slabs of steel fiber reinforced concrete and of plain concrete were made according to the same recipe. For this experiment, high-strength concrete with strength class C70/85 was used. In the mixture of steel fiber reinforced concrete, it was added 30 kg of fibers per 1 m^3, type Dramix RC-65/50-BN. The interior surface of shelters walls must be protected against the concrete spalling, which are caused by detonations on the outside surface. The results of the tests proved that steel fiber reinforced concrete is not a suitable material for the anti-spalling layer.

Keywords: Steel fiber reinforced concrete · Plain concrete · Anti-spalling layer · Ballistic loading · Plastic explosive PLHx 30

1 Introduction

The efficient defense against contemporary threats requires the use of new materials that are able to bear loading from blast, penetration or puncture. The amplifying materials that meet these requirements are described in [1]. In addition, some types of building protection against the explosion of an improvised explosive device are described in [1]. Testing of metallic materials suitable for such protection is described in [2]. The loading structure can also be used for improvised explosives such as disclosed in [3]. This article will also deal with only cement composites.

Concrete, widely used material especially at civil engineering possesses two fundamental disadvantages: low strength at tensile stress and fragility while damaged. Such two disadvantages are partly removed in common constructions by placing in the conventional reinforcement that has been inserted into building blocks as individual steel rods, mesh and/or grids. Both mentioned disadvantages are eliminated using this reinforcement because of rightly placed reinforcement that is capable of to take up resultant force of tensile stresses as well as to avoid the failure of individual fragile construction elements [4–6].

A. Kravcov et al. (eds.), *Durability of Critical Infrastructure, Monitoring and Testing*, Lecture Notes in Mechanical Engineering, DOI 10.1007/978-981-10-3247-9_13

2 Reinforcement

By employing the conventional reinforcement the structure behavior changes, however the concrete properties remain the same. We can reach increase in tensile strength and change of fragile character in concrete failure only by direct reinforcement of concrete structure. To achieve such reinforcement there is a necessity to use reinforcing elements meeting the size of components that create its structure.

The reinforcing element is a fiber. The important advantage of concrete is that all production technologies enable its reinforcement by fibers. Not only fibers influence the properties of fiber reinforced concrete (FRC) but the hardening of FRC in process and finally yet importantly, the properties of FRC mixture of newly made steel fiber reinforced concrete.

From this point of view is the conjunction of fiber reinforced concrete and armored concrete very advantageous. Reinforced concrete itself is resistive nevertheless; its structure remains fragile especially in the armature area.

Presently, there is a research in the area of use of fiber concrete for protective constructions at the department of Engineer Corps under way. As an experimental examination of the dynamic load, the tests of fiber concrete slabs have been carried out with the PlHx 30 plastic explosive.

The search for new materials used for building of protective constructions is a result of development of new weapons that have been used against resistive targets. This is why it is necessary to look for the new materials for new and existing structures as well and adapt such structures to the new requirements [7].

Blast resistance characteristics of concrete with different types of fiber reinforcement are described in [8] and blast impact behavior of concrete with different fiber reinforcement is described in [9].

3 Loading the Test Slabs by PlHx 30 Plastic Explosive Blast

Some test slabs were manufactured of conventional concrete and of fiber concrete, too. To set the guaranteed concrete strength, the test objects for strength tests were being made simultaneously such as 150 mm edge test cubes and for tensile strength tests, the 150 × 150 × 600 mm test squared logs. The test body dimensions were selected with respect to the used Dramix RC-65/50-BN fiber.

The dimension of test slabs was determined with regard to material properties of fiber concrete and slab manipulation while being tested. The tested slabs dimension was 1000 × 1000 × 150 mm and weighted approximately 370 kg, see Fig. 1.

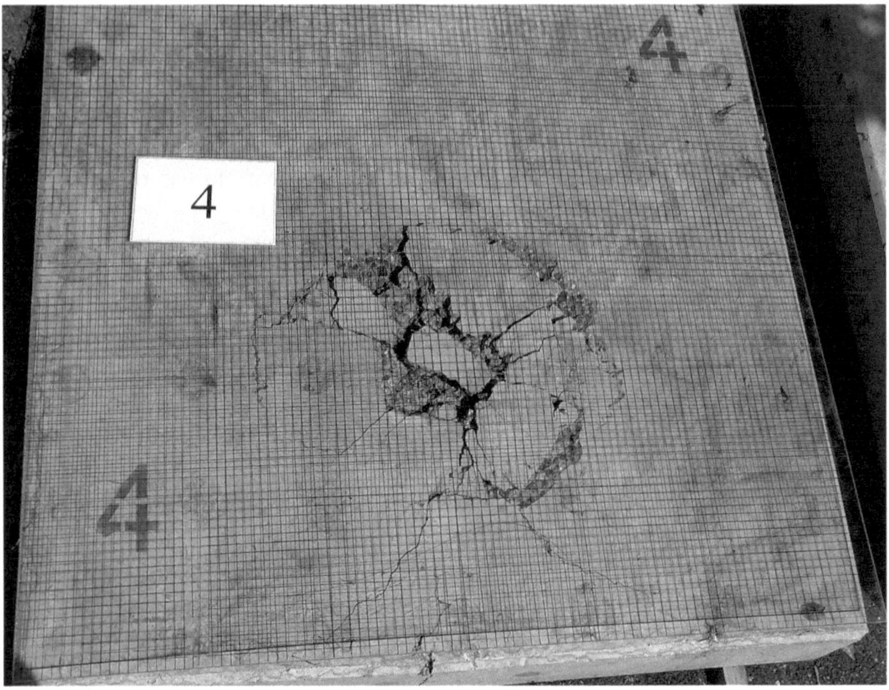

Fig. 1. Tested fiber reinforced concrete slab after being loaded by plastic explosive blast

4 Measuring of Slab Quality by Impulse Ultra Sound Method

The quality of used concrete and fiber concrete was stated prior to dynamic loading by plastic explosive by impulse ultrasound method. The reason to determine quality of material and changes to come after the slab has been loaded by plastic explosive lead to

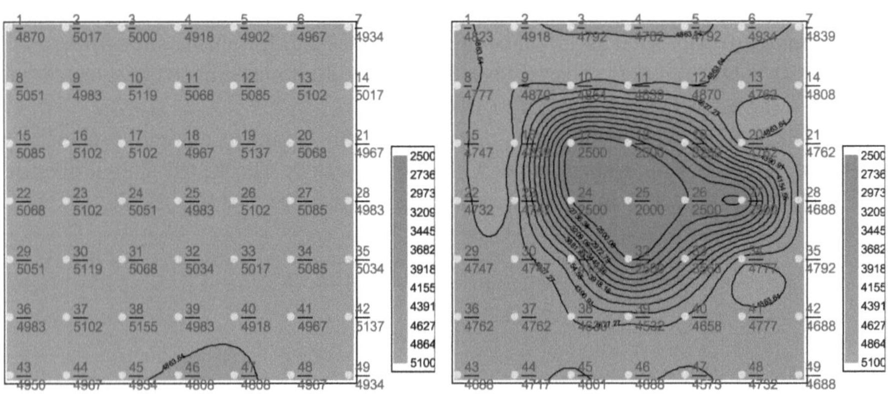

Fig. 2. Layout of the concrete slab izovels prior and after blast loading of 70 g of the plastic explosive PlHx 30. (a) prior (b) after

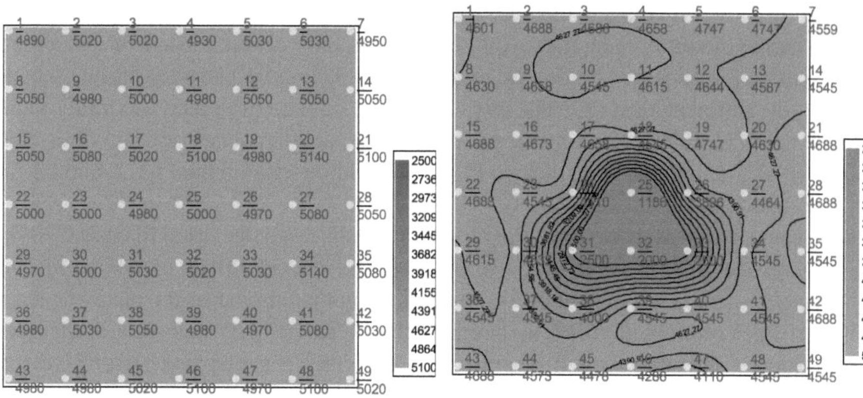

Fig. 3. Layout of the fiber concrete slab izovels before and after blast loading of 70 g of the plastic explosive PlHx 30. (a) prior (b) after

Fig. 4. Tested slab from plain concrete, strength class C70/85, loaded by 70 g of plastic explosive

selection of above-mentioned method. This very method was also employed to learn about dynamic modulus of elasticity of concrete and fiber concrete used. Measurement was taken twice – once just before the initial test of loading by blast and for the second time after the intrinsic blast test. The test was carried out and evaluated according to the standard ČSN 73 1371 [10]. Layout of the concrete slab izovels prior and after blast loading of 70 g of the plastic explosive PlHx 30 could be seen on the Fig. 2, and for the fiber concrete slab on the Fig. 3.

The tested slab was made from the mixture of plain concrete, strength class C70/85, with dimensions 1000 × 1000 × 150 mm, and it was loaded with an explosion of 70 g of plastic explosive, see Fig. 4. For a comparison, another similar tested slab was made from steel fiber reinforced concrete, strength class C80/95, with the same dimensions 1000 × 1000 × 150 mm, and it was also loaded with 70 g of plastic explosive, see Fig. 5. By simply adding 30 kg of steel wires in 1 m^3 of concrete mixture, the ductility of reinforced concrete slabs significantly increased and the slabs showed a greater resistance against contact explosion loading.

Fig. 5. Tested slab from steel fiber reinforced concrete, strength class C80/95, loaded by 70 g of plastic explosive

5 Coefficient of Variation

The coefficient of variation allows you to compare the variability of different selections. It is a relative measure of dispersion (dimensionless) and it is calculated by the following formula [4]:

$$\nu = \frac{\sigma}{\mu} \cdot 100\,[\%] \tag{1}$$

where σ – the standard deviation of the data set for reinforced concrete and concrete separately,

μ – the mean value of all samples.

Variation factor was calculated separately for fiber concrete slabs and concrete slabs Variation factor for fiber concrete slabs:

$$\nu = \frac{\sigma}{\mu} \cdot 100 = \frac{91.98}{4927} \cdot 100 = 1.87\,\%$$

Variation factor for concrete slabs:

$$\nu = \frac{\sigma}{\mu} \cdot 100 = \frac{96.33}{5078} \cdot 100 = 1.90\,\%$$

Considering the results of ultra sound measuring and the result of variation factor (for concrete slabs ν = 1.9%, fiber concrete slabs ν = 1.87%) we can conclude the conventional concrete slabs as well as the fiber concrete slabs were made of high quality material. In respect to the variation factor we can say that despite the fact they were manufactured out of lab they were of laboratory quality (up to ν = 2.5%).

Having the results taken after the loading of slabs by PlHx 30 plastic explosive blast, we could assess what way the specific charge had influenced the existing slab. It was stated there were great differences between the conventional and fiber concrete slabs.

6 Assessment of Spalling

After preparation of slabs, the loading was carried out by plastic explosive blasting. Firstly, the conventional concrete slabs were loaded. After being assessed, the charges for fiber slabs were set. Graphic assessment of relation between the spalling volume and the used explosive is mentioned in Fig. 6.

Distribution of fibers on the spalling surface is somehow equal. The number of fibers matches the size of spalling and indicates the distribution of fibers in the test body. When conventional slabs were loaded by 50 g of plastic explosive, the spalling did not occur. Only some fissures with little fragments appeared on the slab surface. The spalling occurred after the second test was performed. With increased load the volume and depth of spalling rose. Looking at the graph of relation between spalling volume and used explosive, you will find the spalling had risen up to the value of 200 g after the destruction of slab took place two times consecutively.

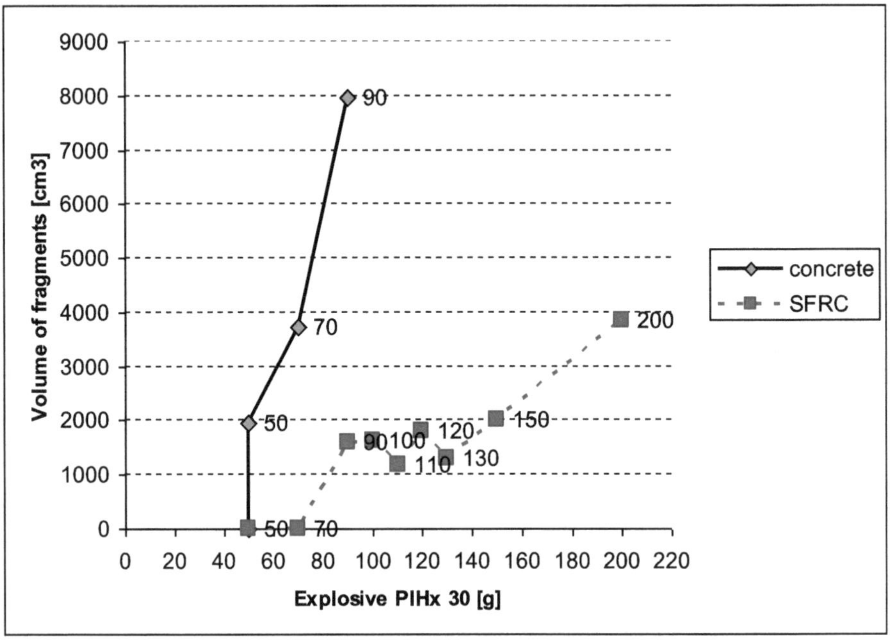

Fig. 6. Relation between spalling volume and use explosive

It also happened that the spalling did not occur in fiber concrete slabs while being loaded by 50 g and next by 70 g of explosive. In the first case, only micro fissures appeared and in the second case the fissures with little fragments showed up. Further, while being loaded the spalling volume had increased rather slowly.

7 Anti-spalling Layer

The inner surface of buildings walls must be prevented from the concrete flaking. This flaking is caused by detonations on the outer face of the material. There is a tensile stress due to the elastic deformation of wave reflection from the inner face of the wall. These tensile stresses exceed the tensile strength of the material and it cause flaking of the concrete. To prevent of the spalling concrete it is essential to adjust the interior walls side by using anti-spalling layer. For this purpose could be used solid or soft anti-spalling layers.

8 Solid Anti-spalling Layer

Steel built into the anti-spalling layer has to be possible, used statically as part of the main bearing function of reinforcement. The Fig. 7 shows example of the anti-spalling layer creation. This anti-spalling layer is called solid because the main load-bearing elements are solid steel beams. Gaps between them are bridged with steel plates and rods ∅ 10 mm and ∅ 16 mm reinforced concrete in the space between the beams.

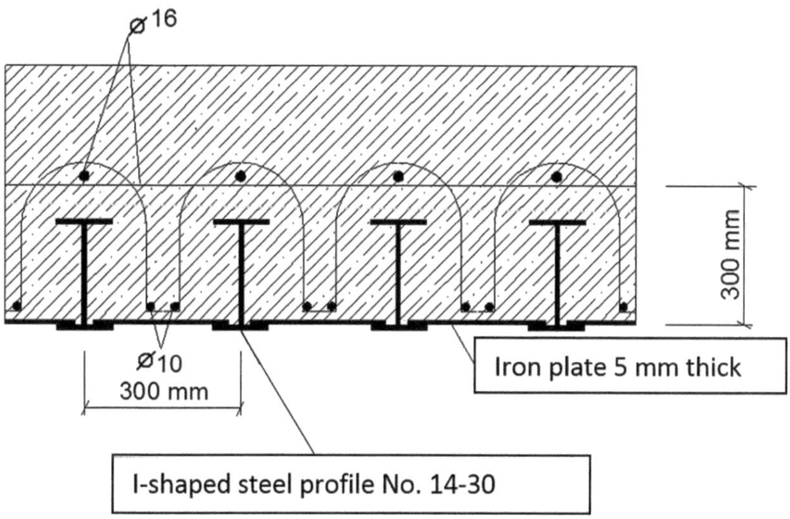

Fig. 7. Solid anti-spalling layer

9 Soft Anti-spalling Layer

Soft anti-spalling layer is created only by concrete reinforcement rods. Usually it consists of several net which is construct from rods Ø 12 mm and Ø16 mm. The reinforcement steel net has square or rectangular meshes. The distance between its rods is 10–50 cm. The densest grid is close to the surface and in the depth of concrete the mesh networks is increasing. Networks must be thoroughly embedded in the inner layers of concrete clips. Soft anti-spalling layer see on the Fig. 8.

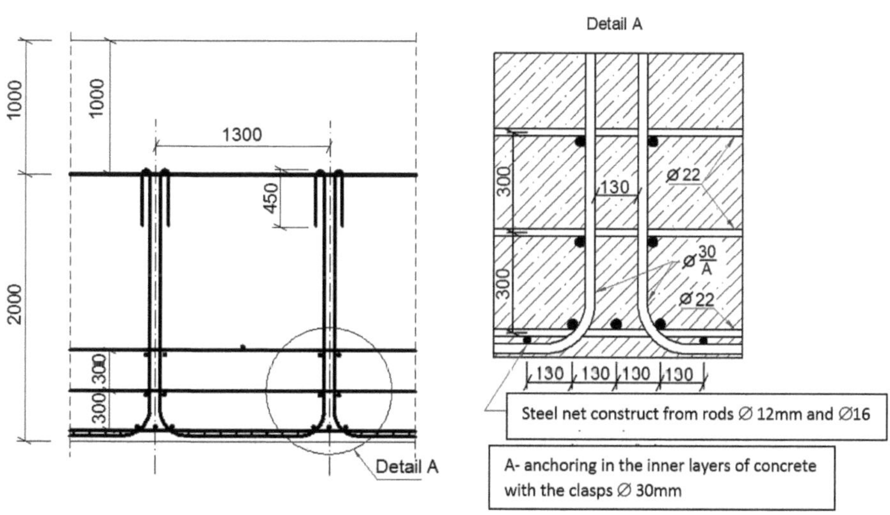

Fig. 8. Soft anti-spalling layer

10 Conclusions

Fiber concrete is one of concrete materials with fiber reinforced structure. It is a concrete that varies from others by structure reinforced by scattered steel fibers. This fact – while being reinforced by certain minimal amount of fibers – is a ground for important change of properties that have been marked as basic drawbacks of conventional concrete. Such reinforcement results in more strength and resistance against dynamic load.

The results revealed the fiber concrete is not suitable for anti-spalling layer. When loaded by blast strength there is considerable tension on inner face that the fiber concrete is not capable to carry, thus resulting in spalling.

The right way to find new materials used for anti-spalling layer could be the sandwich structure. It also applies to armored vehicles where the same effect at protective constructions takes place while being hit by high explosive ammunition.

Well, it is necessary to conclude that fiber concrete probably will not replace the armoring that has been used for reinforcing the protective constructions. However, being compared with conventional concrete the fiber concrete has displayed up to three times higher endurance. Such a fact predetermines afore mentioned material to be used at protective constructions.

More resistant structure (main supporting structure) will be created by incorporation of reinforced concrete and fiber concrete. This way created structure should prove higher endurance against penetration of bombs and projectiles designed to destroy resistant targets.

Acknowledgment. The work presented in this article has been supported by the Czech Republic Ministry of Defence - K 201 Department development program "Development of technologies in weapons construction, ammunition, weapons instrumentation, materials engineering and military infrastructure".

References

1. Figuli, L., Magura, M., Kavický, V., Jangl, Š.: Application of recyclable materials for an increase in building safety against the explosion of an improvised explosive device. Adv. Mater. Res. **1001**, 447–452 (2014). doi:10.4028/www.scientific.net/AMR.1001.447. Trans Tech Publications, Switzerland
2. Figuli, L., Papan, D.: Single degree of freedom analysis of steel beams under blast loading. Appl. Mech. Mater. **617**, 92–95 (2014). doi:10.4028/www.scientific.net/AMM.617.92. Trans Tech Publications, Switzerland
3. Kavicky, V., Figuli, L., Jangl, S., Ligasová, Z.: Analysis of the field test results of ammonium nitrate: fuel oil explosives as improvised explosive device charges. WIT Trans. Built Env. **141**, 297–309 (2014). doi:10.2495/SUSI140261
4. Štoller, J.: The use of fiber reinforced concrete for protective structures. Thesis in the field of "Military construction", Brno, University of Defence – Faculty of Military Technology – Department of Engineer Technology, 254 p (2005)

5. Štoller, J., Zezulová, E.: Testing of critical infrastructure protection against the distant trinitrotoluene blast. In: Proceedings of the 19th International Scientific Conference, Transport Means 2015, Kaunas, Lithuania, Kaunas University of Technology, pp. 505–508 (2015). ISSN 1822-296X
6. Štoller, J., Zezulová, E.: The field testing of high performance fiber reinforced concrete slabs under the TNT load explosion together with the analytical solution and the numerical modelling of those tests results. In: ICMT 2015 - International Conference on Military Technologies 2015, art. no. 7153666 (2015). doi:10.1109/MILTECHS.2015.7153666
7. Maňas, P., Mazal, J.: The reach-back concept in the czech army corps of engineers. In International Conference on Military Technologies, Brno, Faculty of Military Technology, University of Defence (2009). ISBN 978-80-7231-648-9
8. Drdlová, M., Buchar, J., Krátký, J., Řídký, R.: Blast resistance characteristics of concrete with different types of fiber reinforcement. Struct. Concr. **16**(4), 508–517 (2015). doi:10.1002/suco.201400080
9. Drdlová, M., Cechmánek, R., Rídký, R.: Blast impact behaviour of concrete with different fiber reinforcement. In: EPJ Web of Conferences, 94, art. no. 05006 (2015). doi:10.1051/epjconf/20159405006
10. ČSN 73 1371. Ultrasonic pulse method of concrete testing. Publishing house ÚNM, Prague, 20 p (1982)

Analysis of the Different Approaches to Protection of Critical Infrastructures in France and Slovakia

Figuli Lucia[1(✉)], Kavický Vladimír[1], and Picot Sylvain[2]

[1] Faculty of Security Engineering,
University of Žilina, Univerzitná 8215/1, 010 26 Žilina, Slovakia
lucia.figuli@fbi.uniza.sk, kavickya@gmail.com
[2] Ecole Nationale Supérieure de Techniques Avancées de Bretagne
(ENSTA Bretagne), 2 Rue François Verny, 29200 Brest, France
sylvain.picot@ensta-bretagne.org

Abstract. The proposed paper is focused on the different approaches to protection of critical infrastructures. It compares the ways of protection of essential areas for human lives and society in France and Slovakia. It is focused on the Slovak and France legislation about the critical infrastructure, analyses the common features, differences and the methods of protection.

Keywords: Protection · Security · Critical infrastructure · Analysis · Legislation · Sectors

1 Introduction

Security, economic and social stability of the state, its functionality, but also the protection of lives and property of citizens is dependent on the proper functioning of many systems of the state infrastructure. Infrastructure elements necessary to ensure the conditions of life - energy, water, food, communication, transport, health, finance, defence are so closely linked in events which could occur in one of the sectors of infrastructure, they have the potential to cause a shock in the whole social system [1].

The issue of the definition of important infrastructure sectors, so-called Critical infrastructure, its elements and their protection has been emphasized especially after the tragic events from 11th September 2001. Terrorist attacks in the United States confirmed the vulnerability of infrastructure elements and required for new approaches and preparation of sophisticated measures to increase its level of protection [2].

The elements of critical infrastructure are subjected on convention security threat too, as other building used for business or services. There are mainly natural disasters, accidents, catastrophes and crime actions of singles or groups. The situation on the world has required of a protection and defence valuation of critical infrastructure not only on the international level but on the national level too. For these reasons European countries have adopted or are still adopting national standards and national plan for critical infrastructure elements. The present paper is focused on the two European country Slovakia and France. It specifies, analyses and compares the approaches for the critical elements protection in these countries.

A. Kravcov et al. (eds.), *Durability of Critical Infrastructure, Monitoring and Testing*,
Lecture Notes in Mechanical Engineering, DOI 10.1007/978-981-10-3247-9_14

2 Critical Infrastructure

Within the EU member states the European Council Directive 2008/114/EC was given, defined the term of European Critical Infrastructure (EPI). To identify its elements for efficient managing AF ECI security, it is necessary to define the types of attacks as well as the estimate probability of occurrence and their expected consequences. It can be an anthropogenic influence - negative phenomena caused by human activity (e.g. terrorist attacks, hacker attacks, industrial accidents, human errors, etc.) and an natural influences - negative phenomena caused by natural processes (e.g. tectonic effects, meteorological risks related to the effects of extreme weather, risks related to slope movements, etc.) [3].

Dependence of the function of society on function of elements of critical infrastructure, as well as other important objects, makes from these elements a target of different anti – social groups. It is necessary that the area of the protection of such important elements will be provided by trained personnel who can effectively apply technical, technological and regime measures to prevent of treatments accompanied by explosions, and to minimize losses in a crisis situation.

3 Critical Infrastructure in Slovakia

After adopting of documentations of the European Union (for example The Council Directive ES/114/2008) the Slovak Republic has realised fundamental measures for the its critical infrastructure protection. For the codification of the problem of critical infrastructure no existing in laws, its protection was only considered with the connection of the infrastructure defence according to law n. 319/2002 Coll (The defence of the Slovak Republic). However the defence infrastructure does not involve all subjects creating critical infrastructure according this law. It only involves subjects serving for security of the state in war period.

For this reasons the Slovak government has adopted "The Conception of Slovak Critical Infrastructure protection and defence" (called Conception). Goal has been:

(a) To definite the terminology
(b) To sagest sectors of national infrastructure, which are used for the definition of critical infrastructure sectors
(c) To specify criteria for determining of critical infrastructure elements
(d) To specify principles and instruments for critical infrastructure protection and defence elements
(e) To sagest the solution for the protection of classified information (a list of classified information for private sector
(f) To outline of a security research direction
(g) To formulate conclusions and recommendations.

It was suggested 41 sectors and 1 main sector, which were given under the auspices of 14 ministries and central authority of public administration.

The subject with the critical infrastructure protection and defence are considered in Slovakia as follow:

- International partners, international organisations
- The government, public administration
- Regions
- State economical subjects
- Private economic subjects.

Slovakia has adopted the National Programme for the critical infrastructure protection and defence of the Slovak republic (called National Programme) in September 2008 for the creating of condition to guarantee of the security of critical infrastructure.

The goal of the elaboration of National Programme was an evaluation of actual condition and an identification of the most important assets, and the determination of program steps for an improvement of it protection and defence.

It has been identified and fully-fledged 9 major sectors and 14 subsectors of critical infrastructure:

- Water
- Food
- Health
- Energy (subsectors: electro energy, gas production, oil production, mining and metallurgy industry)
- Information and Telecommunication
- Transport (subsectors: road transport, rail transport, air transport and waterways transport)
- Public Policy and Homeland Security (subsectors: Security emergency services – civil emergency services, Fire department and Emergency medical services, Police, Military)
- Industry (chemical, pharmaceutical industry)
- Finance (subsectors: payments systems, accounting system and settlements of transaction by financial instruments).

The ensuring of selected assets critical infrastructure protection had been necessary to lay out in a law.

The law number 45/2001 Coll. has been adopted by National Council of the Slovak republic on February 8th 2001. It had laid out the organisation and the scope of the public administration departments for critical infrastructure, the procedure of critical infrastructure asset selections and the responsibility for disruption.

The essential goal of the critical infrastructure law is the improvement of the present critical infrastructure protection against the increasing threat of terrorist attack.

Considering these facts the law had:

(a) To determine the structure and scope of the public administration departments for critical infrastructure.
(b) To adjust the process of critical infrastructure asset selections. They will be important buildings for the functioning of a society and economy. Some of them, they have cross-border importance (For example the disruption or destruction of them would have significant cross-border impact, outside of the territory of the Slovak republic).

(c) To define tasks of natural and legal persons, which as the operators of critical infrastructure assets, they will be responsible for that.
(d) To determine the sectors of critical infrastructure in which individual elements will be selected.
(e) To solve the administrative-legal relationship in violation of this law.

In this law there is declared the one part of the critical infrastructure system is the defence infrastructure too. It serves for the ensuring of state defence adjusted in the accordance with the special law.

The essential terms are defined in the law:

(a) Critical infrastructure element (named only element) mean mainly engineering building, public service or informational system in the critical information sector. According to sectoral and cross-cutting criteria the disruption or the destruction of them could have negative consequences on the function of society and economy and to life quality of inhabitants from the view of their life protection, health, security, property and environment.
(b) Critical infrastructure sector (named only sector) mean the part of critical infrastructure in which individual elements will be selected. The sector may obtain one or more subsectors of critical infrastructure (named only subsectors).
(c) Critical infrastructure means the system which can be divided in sectors and elements.
(d) Sectoral criteria mean the file of technical or functional criteria with the threshold values, which are used for the element classification for the same sector.
(e) Cross-cutting criteria mean the file of criteria with the threshold values, which are used for the element classification for all sectors.
(f) European critical infrastructure element means mainly engineering building, public service or informational system in the critical information sector. According to European sectoral and cross-cutting criteria the disruption or the destruction of them could have negative consequences on the function of society and economy in Slovak Republic or almost in one Member State from European Union (named relevant Member State).
(g) European sectoral criteria mean the file of technical or functional criteria with the threshold values, which are used for the European element classification for the same sector.
(h) European cross-cutting criteria mean the file of criteria with the threshold values, which are used for the European element classification for all sectors.
(i) Element protection means the ensuring the functionality, continuity and integrity of critical infrastructures in order to order, mitigate and neutralise a threat, risk or vulnerability.
(j) Risk analysis means consideration of relevant threat scenarios, in order to assess the vulnerability and the potential impact or disruption or destruction of critical infrastructure.
(k) Sensitive critical infrastructure information (called only sensitive information) means facts about a critical infrastructure, which if disclosed could be used to plan and act with a view to causing disruption or destruction of critical infrastructure installations.

(l) Operator means a legal person or a natural person who is an owner of elements or is responsible for investments in the element.

State administrative bodies for Critical infrastructure are the Government of Slovak Republic, The Ministry of the Interior or other state administrative bodies which will have responsible for some critical infrastructure sector. They will classify critical infrastructure elements in the critical infrastructure sectors.

The Government of Slovak Republic is required to approve:

- Conception documents of a critical infrastructure development
- Criteria to including and excluding of critical infrastructure elements
- Central register of critical infrastructure elements
- Fulfilling of contact point tasks for the protection of the European critical elements in relation with Member States of European Union and European committee.

Competent Ministries are empowered to carry out the public administration, controls including.

The cross-cutting criteria are provided by the law based on supposed:

(a) Number of threatened person (the number of death and injured person).
(b) Economic influence consisting of:
   1. Economic loss
   2. deterioration in the quality of goods
   3. deterioration in the quality of provided services in the public interest
   4. negative influence on environment.
(c) Influence on inhabitants, it is the disruption of the deterioration in the quality of inhabitants according to:
   1. seriousness of failure of goods supply and its recovery time
   2. seriousness of failure of provided services in the public interest supply and its recovery time
   3. availability of goods supply substitutes
   4. availability of provided services in the public interest supply substitutes.

A condition for an including of critical infrastructure element is a fact of satisfying almost one sectoral criterion and one cross-cutting criterion.

The operator is obligatory to protect against disruption and destruction his or her critical infrastructure element. For fulfilling this task he or she is obligatory:

(a) To implement the security plan, this must be reviewed and updated regularly
(b) To inform his or her employees with the content
(c) To practise at least one or more time a model situation of disruption and destruction element threat according to security plan
(d) To follow the security plan in the moment of disruption and destruction element threat.

Notice: According to law the operator may request financial contribution for fulfilling tasks executing security measures for elements protection (Table 1).

Table 1. Sectors, subsectors of CI elements in Slovakia

Sector	Subsector	Administrative body
1. Transport	Road transport	Ministry of Transport, Construction and Regional Development of the Slovak Republic
	Air transport	
	Waterways transport	
	Rail transport	
2. Electro communications	Satellite communication	
	Networks and services of fixed-line and mobile electronic communications	
3. Energy	Mining	Ministry of Economy of the Slovak Republic
	Electro energy	
	Gas production	
	Oil and oil products production	
4. Information and communication technologies	Informational systems and networks	Ministry of Finance of the Slovak Republic
	Internet	
5. Post	Providing of post services, post payments and procurement activities	Ministry of Transport, Construction and Regional Development of the Slovak Republic
6. Industry	Pharmaceutical industry	Ministry of Economy of the Slovak Republic
	Metallurgical industry	
	Chemical industry	
7. Water and atmosphere	Meteorological service	Ministry of Environment of the Slovak Republic
	Water buildings	
	Drinking water supply	
8. Public health		Ministry of Health of the Slovak Republic

4 Critical Infrastructure in France

Following the attacks of 11 September 2001, France has launched a reflection on the notion of critical infrastructure to modernize the protection of sensitive points and networks. This thinking led to the Decree n°2006 - 212 from 23 February 2006 on the safety of vital activities that are defined as "a set of activities, essential and easily interchangeable or replaceable, contributing to a common goal or to produce and distribute goods or essential services."

Thus, facilities that make an essential contribution to the preservation of war potential and economic security and national survival capacity, or the destruction or damage can present a serious danger to the population, must be protected against threats, especially terrorist.

In 2006 France defined twelve sectors of important vital activities called secteurs d'activités d'importance vitale (SAIV) and it defined operators of vital importance called opérateurs d'importance vitale (OIV), considered as essential for the functionality of the economy and the society. Each sector is attached to a coordinating minister responsible for steering the work and inter-ministerial consultations. 150 vital operators were nominated in seven sectors, including food, water management, energy, health and transport. They began to develop their security plans. The operator designs a two-stage safety system: a safety plan for all of its activities under or treated areas, and protection of individual plans for each of its points of vital importance. A methodological guide was developed for this purpose [4].

The 12 sectors of important vital activity were selected in France:

- Civilian activities of the State;
- Military activities of the State;
- Judicial activities;
- Space and research.

Areas of protection of citizens:

- Health;
- Water Management;
- Food.

Sectors of economic and social life of the nation:

- Energy;
- Electronic communications, audiovisual and information;
- Transport;
- Finance;
- Industry.

The obligations of operators of vital importance are to train managers and board directors of security at both central and local levels. After a risk analysis, to establish an operator security plan taking into account the expected national security directive for

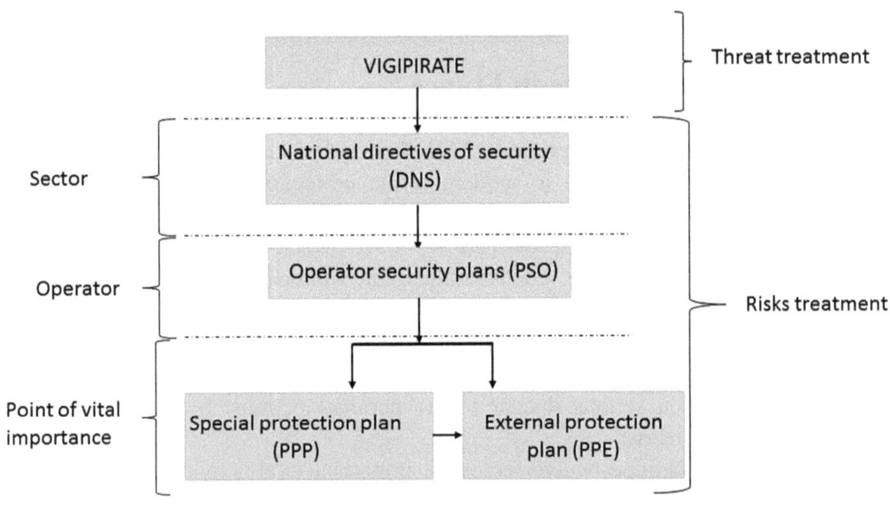

Fig. 1. Hierarchy of the obligations of operators

which they have been designated vital operators. And to identify their points of vital importance which will be a special protection plan (plan particulier de protection - PPP) and an external protection plan (plan de protection externe - EPP) in charge of the department prefect [4] (Fig. 2).

Figure 1 describe the hierarchy of the obligations of operators. So called Vigipirate is one of the tools of the French system of fight against terrorism. This is a permanent device that applies in France and abroad, and all the major segments of society (transport, health, food, energy networks, the computer security …). It aims to involve all national stakeholders that can contribute to vigilance, prevention and protection against terrorist threats: the state, local authorities, operators, citizens.

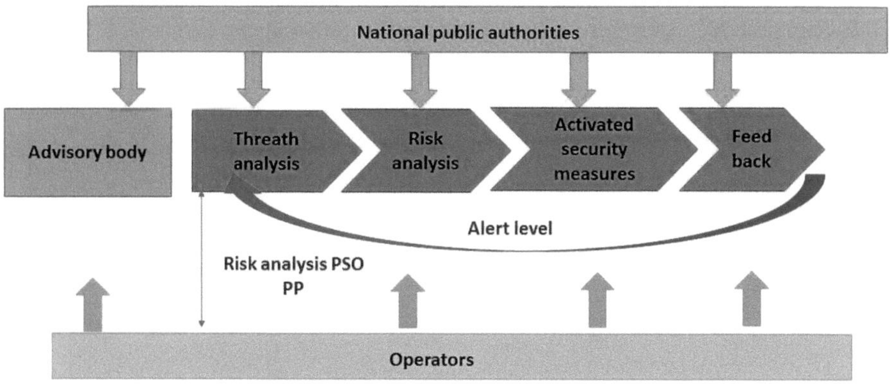

Fig. 2. General organisation according the decrete n°2006 – 212

5 Summary

Following the terrorist attacks of 11 September 2001 in the USA both countries, Slovakia and France, were interested about the security and protection of important subjects for the national economy and society.

Slovakia as well as many European countries adopted European Union Council Directive ES/114/2008 and defined the term of critical infrastructure. Slovakia selected 9 major sectors and 14 subsectors of critical infrastructure to ensure the security of all important areas for the country. France adopted own Decree n°2006 – 212. It not specify the critical infrastructure properly but it speaks about the safety of so called vital activities and defined twelve sectors of important vital activities. The Slovak government set the cross-cutting criteria to including and excluding of critical infrastructure elements. A Slovak operator has to implement the security plan, which must be reviewed and updated regularly. A French operator has to establish security plan after a risk analysis, and to create special protection plan and external protection plan.

Two compared countries proceeded with the protection of essential areas in two different ways. But both are trying to protect what is important and essential – to protect lives, environment and society.

Acknowledgement. This work was supported by the project VEGA 1/0240/15 Process model of critical infrastructure safety and protection in the transportation sector.

References

1. Hofreiter, L., et al.: Ochrana objektov kritickej dopravnej infraštruktúry. University of Žilina/EDIS, Žilina (2013). ISBN 978-557-0803-3
2. Šimák, L., et al.: Ochrana kritickej infraštruktúry v sektore dopravy. University of Žilina/EDIS, Žilina (2012). ISBN 978-80-554-0625-1
3. Dvořák, Z., Leitner, B., Mocova, L.: The need of education in the field of security management and critical infrastructure protection. In: International Scientific Conference – Management 2016, Abstract Proceedings, Belegrade, Mladenovac, Serbia, 5–6 May 2016. Union-Nikola Tesla Univerzity Belgrade, Faculty of Business and Industrial Management, ICIM Plus (2016). ISBN 978-86-6375-053-1
4. http://www.sgdsn.gouv.fr/site_rubrique70.html

Experimental Studies of Blast Pressure Due to Vented Explosion of Methane-Air System

Alexander N. Kravtsov[1(✉)], Jacob Zdebski[2], Vaclav Pospichal[1], and Petr Šelešovský[2]

[1] Faculty of Civil Engineering, Department of Construction Technology, Czech Technical University in Prague, Prague, Czech Republic
kravtale@fsv.cvut.cz

[2] Scientific-Research Coal Institute VVUU a. S., Ostrava, Czech Republic
zdebskij@vvuu.cz

Abstract. A series of tests was conducted to investigate the blast resistances of the building to premixed methane-air interior burning and explosion. Explosion process of the gas methane medium (in the inner space) extremely depend on the object site configuration. An interior deflagration of premixed methane-air burning and explosion in a specific vented chamber can happens due the all hazards: natural, criminal, terrorist, and accidental. The full scale object was tested as control specimen. LVDTs and pressure transducers were used to record deflection histories, and pressure sensors measured airblast pressure histories. In the first stage of a study two experiments of methane-air system in differ proportion were carried out.

Keywords: Vented gas explosion · Shock wave · Methane-air system · Peak pressure · Full-scale experiment · Deflagration

1 Introduction

Today in big cities objects of the civil infrastructure can cause catastrophic damage by sudden change of loading, such as impacts of accidental gas explosion [1], bumps, terrorist attacks [2] or VBIED (vehicle borne improvised explosive devices), etc., and not only substantial material damage can occur but also loss of Human lives can be very extensive. The influence of the shock wave after interior and vented explosion have been studied by experimental tests by Edri et al. in [3, 4], Janovský [5], Walton [6] based on [7] for example, Coltharp and Hall in [8], Pritchard et al. [9], Lebel et al. [10], and many others. Edri et al. [3] studied an interior explosion within a room with limited venting that may occur accidentally in ammunition storage for example. The effect of the pressure distribution on the wall as well as the pressure attenuation depending on these parameters has been gained, and afterburning processes [4]. With using the gas explosion experiments Janovský [5] modeled pressure effects on engineering structures during the accidental explosion of gas in buildings. Saito [11], Mikoid [12] and Nian [13] treated the problems of shock wave by time-dependent numerical methods, such as Godunov [14], multiscale variational method, free hexagons, in combination with the finite element method, finite volumes, boundary elements, etc. are to be applied to solve original mathematical formulations of the problem.

A. Kravcov et al. (eds.), *Durability of Critical Infrastructure, Monitoring and Testing*, Lecture Notes in Mechanical Engineering, DOI 10.1007/978-981-10-3247-9_15

This paper describes the first steps we have taken in developing a multidimensional numerical model to study explosions in large-scale systems containing mixtures of natural gas and air. The aim of this experiments was compare the numerical study of the project design of the structures and norms for building secure and protection with a simulation of the damage effects on the structures based on the field full-scale experiments.

Since the project is concentrated on both theoretical and experimental problems of impacts of gas explosions against solid structures or inside of the structures themselves theoretical part is solved in [15, 16, 17] for example. An interaction problem of the shock wave propagation in combination with finite elements used as a tool for the creation of the matrices of elastic responses of solid structures (elastic arch) is solved in [15]. The algorithm is based on Godunov's scheme [14] involving deformed finite difference meshes modified to increase the time and space parameters. A variational approach is adopted to generate the moving finite difference mesh. Based on the previously mentioned, the method is applied to the description of the response of a tunnel to blast pressure in [16]. The impact of the shock wave on the elastic structure embodies pressure on the structure, reflective waves causing wave interaction.

2 Experimental Part

Field experiments were designed for studying the blast load after the vented gas explosion of a methane-air system. The blast tests were performed in a structure, specially selected for this series of tests. This structure was perfectly fit for the burning and explosion of the homogenized methane-air mixture and for recording deflection histories and measuring over-pressures thanks to a different space configuration and dimensions of the structure's chambers. The first phase of the series of tests was conducted in structure one. The structure's parameters are presented in Figs. 1, 2 and Table 1.

3 Experimental Objectives of the Test

- To create an explosive mixture, natural gas was infused into the rooms from pressure cylinders containing 92% of methane
- The rooms were without any built-in systems
- Two SO instant detonators located by the wall of the room were used for the initiation
- The methane-air mixture was homogenized using explosive proof electric fans with a diameter of 400 mm from the bottom point to the top diagonally
- Continuous sampling of the methane concentration in the room was provided in two places in 1/3 and 2/3 of the room height, the sampling tubes for the measuring of the methane concentration had been installed in places envisaged for forced sampling
- Ventilation holes were covered with PE-foil and all the leaks in the walls and the ceiling were sealed by mounting foam
- Over-pressures in the room were measured at 6 places (4 in the fumigated space and 2 out of it)

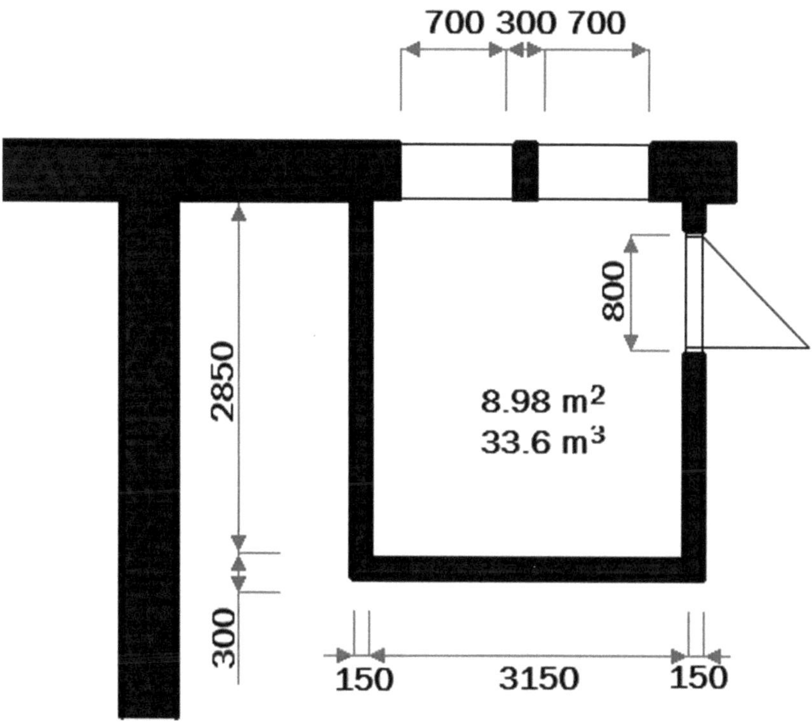

Fig. 1. Plan view of the structure

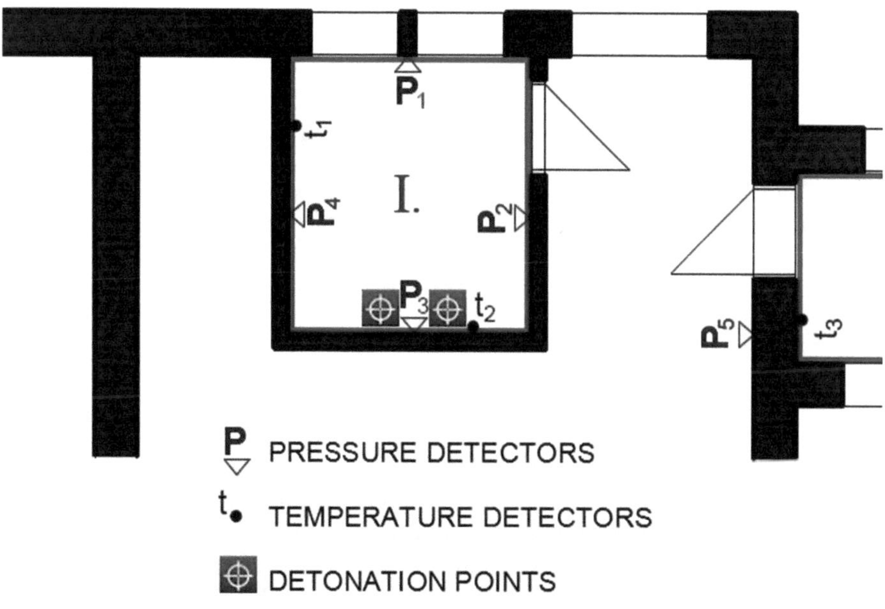

Fig. 2. Scheme of the pressure and temperature sensors location

Table 1. Structure one dimensions

Room #	Height [m]	Area [m^2]	Volume [m^3]	Openings [m^2]
I.	3,75	9,0	33,6	1,6

- Temperatures were measured at two points inside the fumigated space at a height of 1.5 m
- Natural gas was fed via the suction part of the fan and dispersed into the space.

Interior explosions of a gas system (in our case methane or natural gas) in the structures of building infrastructure mostly have the nature of vented gas explosions (windows, doors, lightweight surfaces) and, unlike fully enclosed explosions, the generated pressure depends on the volume, the geometry of constraints, inside blocking and the burning velocity. It should be noted that under these conditions, explosion phenomena with the nature of deflagration burning with pressure effects arise. The ideal pattern of a vented explosion has two pressure maxima, the first being characterized by the development of an explosion and an increase in pressure over time, followed by the removal of the valve ("opening" of the structurally weakest member) with a pressure drop due to ventilation; subsequently, the maximum surface area of flames is reached (together with the second pressure maximum), being followed by the ventilation of the pressure of combustion products and a sharp rise in temperature. The absolute magnitudes and differences between both pressures depend, among others, on the size of the area and the resistance of material (adequately the area and weight of the valve).

Based on the estimation of the parameters of vented explosions made by Janovský [17], according to [18, 19], the following equations very well comply with the values of the first and the second pressure P1 and P2:

$$P_1 = S_0 \cdot \frac{(4,3 \cdot K \cdot W + 28)}{V^{\frac{1}{3}}} \tag{1}$$

$$P_2 = 58 \cdot S_0 \cdot K \tag{2}$$

where P_1 a P_2 are in millibars, S_0 is the laminar burning velocity of gas in m · s^{-1}, V is the volume of the enclosed space in m^3 and W is the basis weight of the valve in kg · m^{-2}.

The coefficient K is the ratio of the area of the wall in which a hole is made to the area of the valve:

$$K = \frac{A_s}{A_v} \tag{3}$$

If there are constraints (obstacles) in the room, S0 is multiplied by the coefficient of 1–5, which expresses the turbulation acceleration of the explosion.

As we are professionally engaged in experimental drifts in Štramberk, mainly in the area of 1D or 2D propagation of explosion waves (having specific experience with vented explosions, see [5], we tried to predict the expected maximum pressure according to the relations formulated by Cubbage and Simmonds (1) and (2) whose

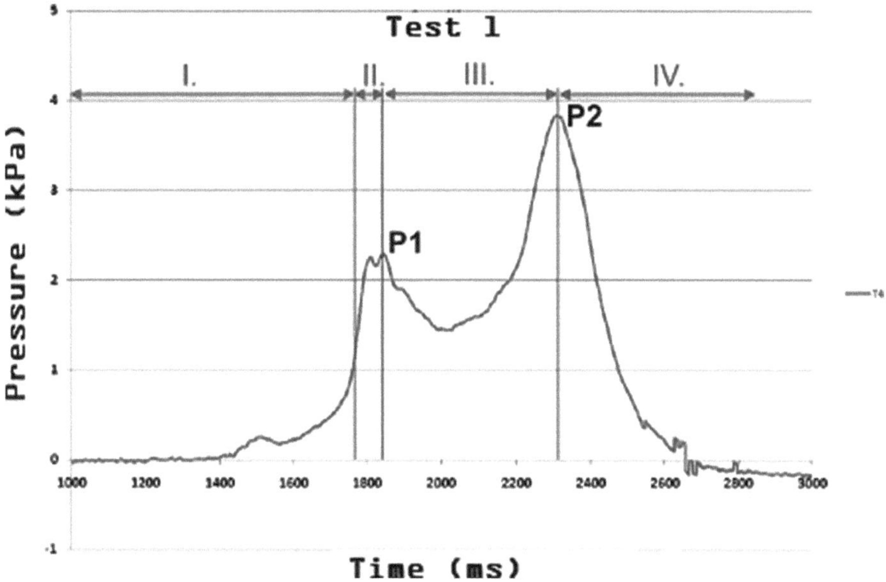

Fig. 3. Distribution of vented explosion

conditions for use were generally met by the first room. The maximum value of the second pressure maximum could reach up to 17.45 kPa, while for the first pressure the maximum was 7–9 kPa (the value *W* was set by a technical estimate). Since the initiation was not assumed to occur in the centre of mass of the room, but by the wall, lower actual pressures could be expected. The explosion pressure pattern and its effects in experimental measurements depend primarily on the volume of an explosive system, interior built-in systems and related turbulence, the geometry of constraints, the burning velocity, the homogenization state of the mixture and the concentration of the explosive system. A characteristic pattern of an interior vented explosion with the nature of deflagration burning with pressure effects may theoretically be phased as follows.

I. In the first stage of explosion – development of the shock wave and the gas dynamical streaming in time is recording. Ventilation opening is still clouse.
II. In the second stage vents are opened – first pressure P1 is measured in time when the shock wave hits the most weakest structural element.
III. In the third stage Fáze ventilování – dochází k částečnému uvolnění a poklesu tlaku.
IV. In the last stage the maximum of fire font and venting of the compaction products – the second pressure maximum P2 is measurede. Major measured pressure parameters are presented in Table 2.

It must be added that each experimental test was unique and specific by its geometrical setup, the layout of walls, vents and related pressure patterns. If we concentrate on tests No. 1 and 2, which took place in the smallest room I (see Figs. 1, 2 and Table 1), we may state that they represent a classic interior vented explosion. The

Table 2. Measured parameters of the test's 1 and 2

Maximum gas pressure by a deflagration of the Methane-air system														
Test #	CH_4 [%]	Gauge/ Pressure [kPa]												t^+ [s]
		T1		T2		T3		T4		T5*		T6*		
		p_1	p_2	p_1	p_2	p_1	p_2	p_1	p_2	p_1	p_2	p_1	p_2	
1	7	2,56	4,07	2,16	3,72	2,20	3,30	2,50	3,83	1,33	*	0,20	*	1,0
2	8–9	3,02	3,29	2,95	2,90	3,03	2,73	3,00	3,15	1,49	*	0,21	*	1,0

Where $\mathbf{t^+}$ - is a time of overpressured stage
Poznámka:* detectors out of methane-air system in open space
** detectors out of methane-air system, but in zone of explosion
*** only one maximum pressure

pressure patterns are illustrated in Figs. 4 and 5. Glass bricks were built in the room and the door space was sealed by a double PE foil.

In test No. 1, the explosion was vented and, as indicated in Fig. 3, there was an almost ideal interior vented explosion with two maxima. Except for a few tiny cracks in the masonry, the room remained intact after the test, and, at a lower concentration of methane (7%), the vent managed to protect the interior masonry from more extensive damage.

In test No. 2, the methane concentration had been increased to approach stoichiometric ratios (8–9%). Following the initiation of the above methane mixture, the first pressure maximum P1 increased compared to test 1, and the weakest brick partition by the door vent (15 cm in thickness) was partially disintegrated and collapsed,

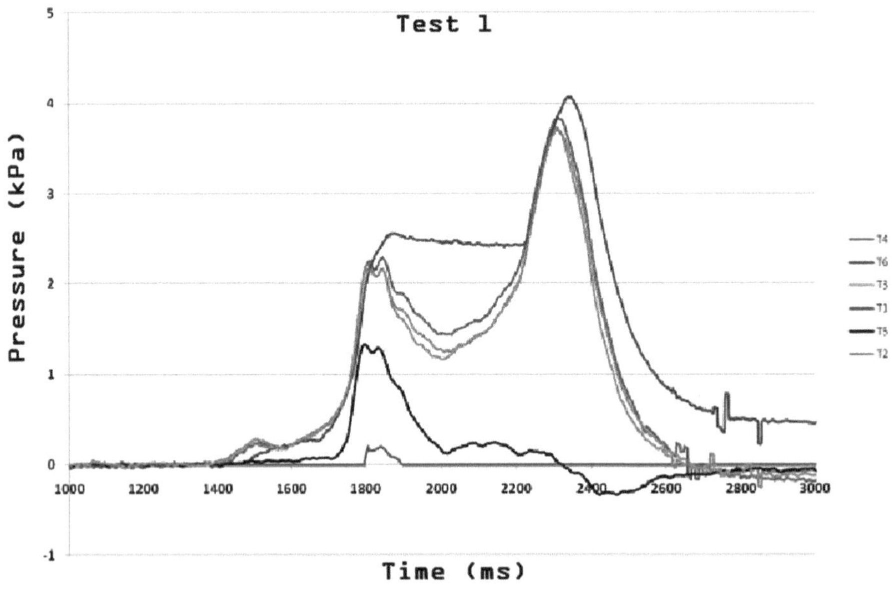

Fig. 4. Pressure-time history for the 1st test.

thus releasing the pressure outside the room. For this reason, the second pressure maximum P2 no longer reached the values of the first test. The collapse and subsequent cracks in the masonry are documented in Figs. 6 and 7.

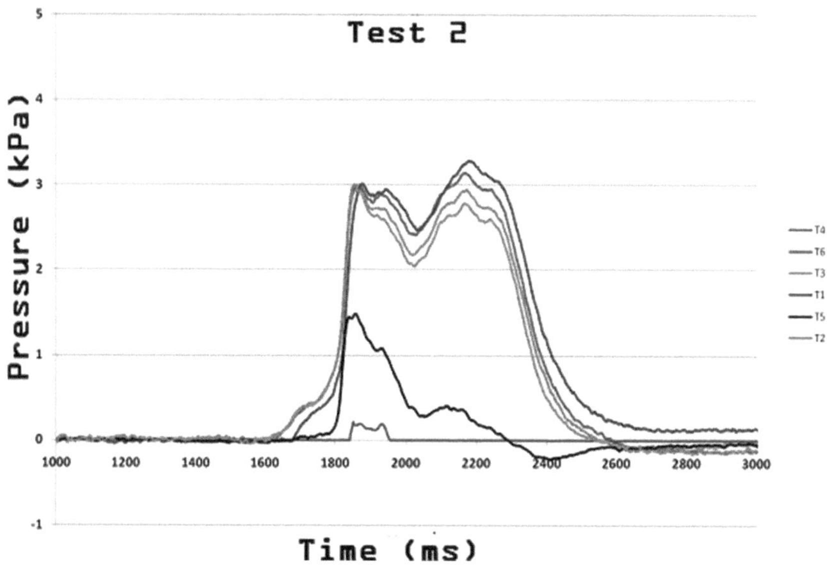

Fig. 5. Pressure-time history for the 2nd test.

Fig. 6. Demolition of the wall

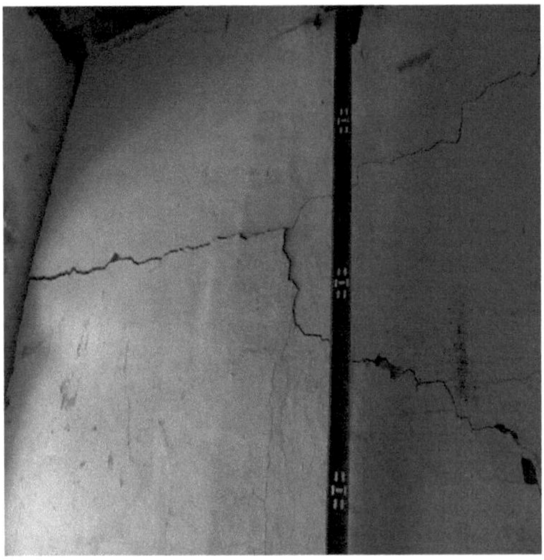

Fig. 7. Cracks in the masonry

4 Conclusions

Interior vented explosions of flammable air-gas mixtures in overground buildings always highly depend on the building's shape, the size of the explosion space, the built-in system's workmanship, etc. As confirmed by the above-described experimental tests, it is also the size and the basis weight of the vents that plays an important role of consequences after the ignition of natural gas. Here, the important aspect is how long and at what pressure the initially enclosed interior space of a fumigated room is connected with the outside environment. In this way, a fresh, unreacted mixture or fumes may escape out of the enclosed space. It is obvious that the larger the area of the vents, the more dramatic the reduction in the explosion pressure and the subsequent destructive effects. These processes are also accompanied by a significant increase in temperature inside the fumigated room over a short time. The maximum temperatures measured at a height of 1.5 m ranged around 1000 °C.

It is also necessary to take into account the fact that release ventilation design elements may represent a serious hazard for the surrounding area during an explosion. Lighter fragmentation with distinct thermal radiation may also cause subsequent damage. Emergency situations arising due to the escape of methane into buildings manifest that methane layers are predominantly produced as a result of a low effective ventilation of the spaces. Although the explosion parameters are of lower intensity during the initiation, the period of thermal radiation into the surrounding area gets considerably longer, which raises the risks of subsequent fires. These emergency situations cause not only financial losses to property, but also pose a serious threat to human health and life.

Acknowledgments. Financial supports for this work, provided by the Ministry of the Interior of the Czech Republic (MV VG20102013055) and the Czech National Ministry of Education (No. 50534090) are gratefully acknowledged. The authors are very grateful to colleague Ing. Petr Šelešovský, who also played major roles in this collaboration with the field experiments.

References

1. Hirano, T.: Accidental explosions of semiconductor manufacturing gases in Japan. J. Loss Prev. Process Ind. **17**, 29–34 (2004)
2. Gebbeken, N., Döge, T.: Explosion protection: architectural design, urban planning and landscape planning. Int. J. Protect. Struct. **1**(1), 1–21 (2010)
3. Edri, I., Feldgun, V.R., Karinski, Y.S., Yankelevsky, D.Z.: On blast pressure analysis due to a partially confined explosion: 1. Experimental studies. Int. J. Protect. Struct. **2**(1) (2011)
4. Edri, I., Feldgun, V.R., Karinski, Y.S., Yankelevsky, D.Z.: Afterburning aspects in an internal TNT explosion. Int. J. Protect. Struct. **4**(1), 97–116 (2013)
5. Janovský, B., Šelešovský, P., Horkel, J., Vejs, L.: Vented confined explosions in Stramberk experimental mine and AutoReaGas simulation. Loss Prev. Process Ind. **19**, 280–287 (2006)
6. Walton, B.: Designing blast hardened structures for military and civilian use. PE, Protective Design Center, US Army Corps of Engineers, Omaha, NE. The AMPTIAC Quarterly, vol. 6, no. 4 (2005)
7. US Army Corps of Engineers, Naval Facilities Engineering Command, Air Force Civil Engineer Support Agency. Design of structures to resist the effects of accidental explosions. UFC 3-340-02, US Department of the Army and Defense Special Weapons Agency, Washington, DC (2008)
8. Coltharp, D., Hall, R.L.: Blast retrofit research and development: protection for walls and windows. Geotechnical and Structures Laboratory, US Army Corps of Engineers, Engineer Research and Development Center, Vicksburg, MS. The AMPTIAC Quarterly, vol. 6, no. 4 (2005)
9. Pritchard, D.K., Freeman, D.J., Guilbert, P.W.: Prediction of explosion pressures in confined spaces. J. Loss Prev. Process Ind. **9**(3), 205–215 (1996)
10. Lebel, L.S., Brousseau, P., Erhardt, L., Andrews, W.S.: Thermochemistry of the combustion of gas phase and condensed phase detonation products in an explosive fireball. Combust. Flame (2013). http://dx.doi.org/10.1016/j.combustflame.2013.10.024
11. Saito, T., Yamashita, H., Takayama, K.: CFD application to construction of hazard maps of volcanic eruptions. In: ASME Pressure Vessels and Piping Conference, Canada (2002)
12. Mukoid, V.P.: Transient dynamic behavior of a closed gas-filled shell subjected to internal blast loading. Int. Appl. Mech. **35**(3), 288–294 (1999)
13. Nian, W., Subramaniam, K.V.L., Andreopoulos, Y.: Dynamic compaction of foam under blast loading considering fluid-structure interaction effects. Int. J. Impact Eng. **50**, 29–39 (2012)
14. Godunov, S.K., Zabrodin, A.V., et.al: Numerical Solutions of the Poly-dimensional Problem of Gas Dynamic. Nauka, Moscow (1976) (in Russian)
15. Kravtsov, A.N., Zdebski, J., Svoboda, P., Pospichal, V.: Numerical analysis of explosion to deflagration process due to methane gas explosion in underground structures. In: ICMT 2015 - International Conference on Military Technologies 2015, 9 July 2015. Article number 7153696

16. Kravtsov, A., Svoboda, P.: Experimental studies of the blast pressure due to an explosion in the tunnel. In: 6th International Symposium on Tunnel Safety and Security, Marseille, France (2014)
17. Janovský, B: Použití tabelovaných hodnot vlastností pevných a plynných výbušin pro odhad zatížení. Vzdělávací seminář – pilotní kurs projektu "Protivýbuchová ochrana staveb" - Odezva stavebních konstrukcí při zatížení výbuchem a jejich ochrana. ČVÚT v Praze (2008). ISBN 978-80-01-03991-5
18. Cubbage, P.A., Simmonds, W.A.: An investigation of explosion reliefs for industrial drying ovens – I top reliefs in box ovens. Trans. Inst. Gas Eng. **105**, 470 (1955)
19. Cubbage, P.A., Simmonds, W.A.: An investigation of explosion reliefs for industrial drying ovens – II back reliefs in box ovens. Trans. Inst. Gas Eng. **107** (1957)

Theoretical Aspects of Critical Infrastructure Protection

Ladislav Hofreiter(✉) and Zuzana Zvaková

Faculty of Security Engineering, University of Žilina,
Univerzitná 8215/1, 010 26 Žilina, Slovak Republic
{Ladislav.Hofreiter,Zuzana.Zvakova}@fbi.uniza.sk

Abstract. Theoretical, especially ontological approach to security examining in sector of critical, infrastructure is associated with a clear description of the object of interest, the expression of the principle of ontological security. Object of interest is the critical infrastructure sector, systematically organized into sub-elements and, as well as clarifying what the protection security is, what are its attributes and factors. The subject of research will be also interactions between security factors and their impact on the security situation. Clarification of ontological principles is to identify and to describe the causes that allow or give the security reason for existence, therefore something exists or may exist.

Keywords: Critical infrastructure · Sector · Object of interests · Security · Protection

1 Introduction

Security, the economic and social stability of the country, its functionality also protection of life and property of citizens depends on the proper functioning of many infrastructure systems of the country. The infrastructure necessary for life - energy, water, food, communications, transport, health service, finances, defence - are closely related. The events that may arise in one of the sectors of infrastructure (in one of the subsystems of the society) are able to disrupt all society systems.

We call infrastructure as a critical infrastructure if disruption of function, lack or destruction of physical and virtual systems, institutes, equipment and others services could cause disruption of social stability and state security, create crisis situation or seriously influence the function of state administration and autonomy in crisis situations. This definition of critical infrastructure expresses contents and meaning not only for the functioning of state in normal conditions but in crisis situations too. Crisis situations could be initiated knowingly or unknowingly and caused by internal or external factors or natural, technology and social factors.

The system structure of the critical infrastructure is defined in the relevant documents as "system, which is divided into sectors/subsectors and elements" [1]. The object of our research is the protection and security of critical infrastructure, with relevant subsectors and elements.

A. Kravcov et al. (eds.), *Durability of Critical Infrastructure, Monitoring and Testing*,
Lecture Notes in Mechanical Engineering, DOI 10.1007/978-981-10-3247-9_16

2 Ontological Aspects of the Examined Critical Infrastructure Protection

Ontology (gr. Ontos - being, gr. Logos - learning) is the most commonly comprehended as "the doctrine of being". This is the approach of philosophers and philosophy but not only have they used this concept. In various fields of human activity, including security research is for universal, universally understandable and graspable way need to know how to describe reality, as well as to define real objects, their attributes, and relationships among these objects. Research in various fields requires describing what is, what exists, what can be or what is able to exist. Is it possible to use an ontological approach. The term "being" refers to real, concrete existing things, in this approach, but it can refers to things that may exist too.

Ontological approach in examination of CI allows:

- to describe uniquely the object of interest,
- to express ontological principle of protection of CI.

Description of the object of interest is based on the necessity to express the subject of examination. This is what interests us, what we want to clarify and what we want to solve. The object of interest, respectively a research problem is obtained selecting problems existing in the reality. This process could be presented at the model of selection of object interest in the protection of critical infrastructure in the particular sector (Fig. 1).

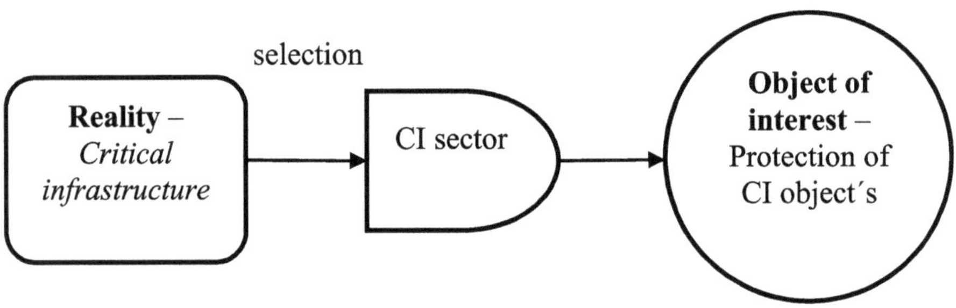

Fig. 1. Selection of object of interest in reality

Being is the reality for us – it is the form of existence of systemically arranged the sector. We selected CI of particular sector from this reality. Protection of its elements is the object of interest. The object of critical infrastructure becomes the object of protection. The object of protection may be a physical, a tangible object or an intangible asset. We need to know, if is necessary to protect a material object or group of objects, or if we need to protect the features performing or providing the objects.

The second aspect is about the terms of ontological principle in CI protection. The essence of this principle is the identification and the description of causes and to give importance of being to protection. That means, the protection is something what exists or may exist. In the case of the critical infrastructure may be more principles.

Ontological principles for CI protection:

- the existence of CI elements, having existential meaning and importance for the function of the CI sector/subsector,
- the existence of real or potential threats, which have may an affect the functionality of the CI elements. This effect may result in a disruption in the CI sector or whole national infrastructure.

Protection system may exist without the second ontological principle. Protection systems are designed mainly to prevent in the practice and resource for discarding potential offenders. The protection system is needed if it required from the importance of element. Threat's may be latent in this case.

3 The Content and Meaning of the Protection

Ontological approach to protection means to clarify what is protection, how their attributes are, what structure it could exist in and how relations are in structure of protection. The term of protection and its importance is the fundamental research problem.

The term protection may be comprehended in various meaning. The most general meaning is a care about averting risk and different harmful effects from the environment of social subject or material object, that may threating its safety. Protection is "summary of system measures, activities and means to prevent *and* eliminate of consequences of current and potential internal and external threats…" according to the Terminological dictionary of crisis management [2].

Protection is summary of measures to divert or eliminate of harmful effects and consequences of incidents and crisis situations according to the Explanatory Dictionary of Security Management. The protection is also *act of protecting or existence of protection* or as *safeguard - something that guarantees security* in this dictionary [3].

There is term *protection of property* in this dictionary and it is defined as *a summary of safety, technical and regime measures which lead to obstruction of enemy action against the object and people in the object. The goal of these measures is to prevent of enemy action against people and property* [3].

Protection is *"to create of the safety environment for the subject" written* by *Lukáš* and it is realized designing all available means for ensuring required or defined safety [4].

The subject of protection may be tangible or intangible property, important objects, spaces or people.

Protection of spaces or objects is *summary of measures and activities for prevention of events, acts or phenomenon that could threat protected objects or spaces as well as to prevent entering unauthorized people to protecting objects or spaces.*

We can examine protection in two basic meaning in terms of presented view:

- Protection as an activity for ensure safety
- Protection as means or system of means for ensures safety.

Protection as an Activity. Protection as activity is to create safety environment for CI objects. This activity is focused on:

- elimination of negative effect - social, natural and technogenic threats on safety of objects,
- reduction of risks resulting from the environment to acceptable level,
- ensuring the functionality and operability of the system for solving crisis situations in the environment of the object.

Critical Infrastructure protection is defined as the set of measures planned and performed with the goal:

- Critical Infrastructure protection is defined as the set of measures planned and performed with the goal:
- Identify and protect critical objects from sectors of state infrastructure. It could be critical from some aspects – maintenance its safety, functionality, economics and society stability. It is necessary for equivalence review in state and private sphere.
- Ensure functionality of early warning system about crisis situations and ensure protection infrastructure important for solving crisis situations [5].

Protection of CI is Risk management method. It means it is a realization of set coordinated and systematic activities. These activities are focused on the elimination of the possibility of security threat intention (ensure safety CI objects) by a negative event, effect or activity.

Protection of CI objects is a systematic activity for ensuring safety using protective equipment and measures. These equipment and measures are for ensuring functionality, integrity, and continuity of activity object. The aim is to prevent, to avert or to ease threat of object or its destruction (modified by [1]).

Protection as a Means. Protection in closer meaning is a mean (individual, technical equipment…) *offered to protect* another person, material values, CI objects etc.

Means for objects protecting, spaces and assets of different kind could be divided into:

- *Passive protection elements* (mechanical barrier, building elements, barriers, foils etc.),
- *Active protection elements* – alarm systems (electrical barrier, CCTV, entry control system, fire detection)
- *Physical protection elements* (security service, self- protection)
- *Regime measures* – preparation and creating procedures for application protection system.

The protection system is protection means connection. This system has various definitions in various spaces.

Physical Protection System it is a system that integrates people, means and procedures for property protection or object protection against theft, sabotage or other human attack [6].

Система физической защиты in the Russian speaking area. It is a set of laws, organization measures and technical solutions for protection of important interests and resources in business. Protection is focused on human attacks threats [7].

We understand *Protection System* in connection with the solved problem as *integrate set of real elements, actions and processes which are logical and functionally arranged for the instrument for ensure CI protection in the time and space. It may be from aspect of systemic approach synergistic system with the target behavior.* Required *functions* of protecting system are:

- *deterrence of* potential attackers from the attack to protected object,
- *detection* of intrusion to protected object, spaces or zones or detection of dangerous situation formation in object or in its close environs,
- *delay* of attackers movement,
- *response* on disruption of protected spaces or zone, to the disallow attackers approach to protected interest.

Protection is connected with the *ensuring of safety process* in previous meanings. This process is to ensure safety for someone or something. It is the primary goal of responsible authorities and institutes in CI protection. This is the reason to know and understand the term of security.

4 Security Assessment of Objects of Critical Infrastructure

Level of security of objects can be rated. Base for this rate are following factors [8]:

- *Vulnerability of* quantified defensive and protective skills.
- *Threat* which may be in open or latent form in environment.

We can express it by relation:

$$S(t) = f\{T(t); V(t)\} \quad (1)$$

S(t) *is a real security of object in time t,*
T(t) *is an intensity of the threat in time t,*
V(t) *is a vulnerability of object against real threats in time t.*

Security Assessment. *The threat* is the term for a particular, physically existing object, effect or event. This particular, physically existing object, effect or event is able to cause damage or injury. It is everything, what is dangerous for an object and could negatively change its security situation, in the widest meaning. The threats are events and effects, which could happen in relative short time or they happened in the past and could cause dramatic changes to objects in existential conditions too. The threats occurrence is determined by:

- the occurrence of the wearer, the source of the threat - something or someone which could cause damage or injury or who intends to do it.
- the capacity of the wearer, the source of the threat endanger the subject or protected interest or causes damage, injury or dangerous event.

Necessary but not enough conditions for the real threat are occurrence of *the source* or wearer of the threat, which *wants* (has a motivation) and *knows* (has the capacity) for the threat. These conditions do not guarantee the real threat. There are *other conditions* for real and direct threat, for example, the adequate vulnerability of the object, availability of protected interest, low risk of failure for attackers, high probability of profit and others favorable conditions for attackers from the inside of the object etc.

The size of every identified threat can be rated using qualitative method with expert review of answers (Fig. 2) for the following questions:

- Does a source of danger exist (S – *Source*)?
- Has a reason or motivation (M– *Motivation*)?
- Has a capability for attack (C – *Capability*)?
- Does event occur in past (E – *Event*)?

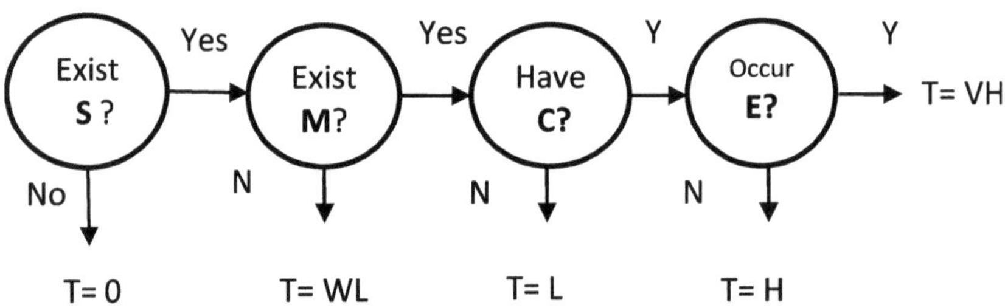

Fig. 2. Model of threat assessment

Vulnerability Assessment. Vulnerability is an attribute of any material objects or technical means to lose ability to do its natural or given function. It is due to action internal or external threats of different character and intensity. It is degree or the extent of material objects or technical means ability to resist internal and external threats.

Vulnerability is a part of the protected object (building, fenestration) or the elements of protection system which do not provide required protection degree (protection equivalent to the security class), are weak or easy crossable element in protecting the system or create appropriate conditions for attacking on an object, increase attack probability and its success.

Indicators for vulnerability rating of the object, its elements, zones or spaces are:

- *Level of vulnerability* rated by qualitative methods (expert review) as little, medium, large etc.,
- or *probability of successful attack* on the object P_A, it can be expressed with using probability models in the scale (0;1), or like subjective probability, respectively credibility, or according to the relation:

$$P_A = 1 - P_E \tag{2}$$

P_A is probability of successful attack,
P_E is probability of attack elimination; it is express according to the relation (3)

$$P_E = f(P_{d,l}, P_{I,L}) \tag{3}$$

P_E is probability of attack elimination
$P_{d,l}$ is probability of detection attack to critical line of detection[1],
$P_{I,L}$ is probability of interruption of attack before line areas endangered of object[2]

Security Assessment of the Object. Determining of the CI object security lines, in the logical analysis of the relationship of factors and their impact on the object security.

- If size of identified threat is reviewed as a large but vulnerability is reviewed as very small; it shows to enough security, also for the protection of the element. It is because of the threat cannot accord to degree of protection system resistance cause damage or disruption to CI objects functions. The method is illustrated in the Table 1.

Table 1. Security evaluation matrix.

		Threat				
		Very low (1)	Low (2)	Not low (3)	High (4)	Very high (5)
Vulnerability	Very low (1)	1	2	3	4	5
	Low (2)	2	4	6	8	10
	Not low (3)	3	6	9	12	15
	High (4)	4	8	12	16	20
	Very high (5)	5	10	15	20	25

Legend: 1–5: sufficient security; 6–10: threatened security, the risk situation; 12–25: not enough security or no security

This method and scale of evaluation is only for example. The evaluator may determine own scale and method for evaluation according to extent of acceptable risk.

---

[1] Critical line of detection is the latest option detection disruption to effective intervention of units of physical protection (Intervention Unit).

[2] Line of areas endangered of object is distance from limit of the object, from which the object or protected interest can endanger. If is protected interest in the object, is this limit determined by inside arrangement protected areas.

5 Conclusion

The more secure Critical infrastructure object will be, the higher *abilities of protection subjects to* identify possible threats in time will be. And its *possibilities for permanent protection* of its physical integrity and fulfilling its functions against changing (evolving) security threats of various kinds will be. However, this is expected active approach of the object. It means especially:

- to identify possible threats, its resources and holders,
- to detect of proximate (immediate) causes of security threats; it means to identify why may be threatened object security,
- to detect ultimate (final) causes of security threat, i.e. how can reach to object security threat [8].

The situation when the object will be secure, does not mean that it exists without threats in environment. Security means that the object is enough resistant against all expected negative effect and events which may change conditions of its stabile functioning.

The conditions for object security are:

- there is not threat source, there is not threats itself or its environment (others systems, processes and objects),
- its condition guarantees stability, physical integrity or continuity of fulfilment its functions,
- protection system of the object has got enough potential to eliminate or minimize external or internal various kind of threats,
- it is able immediately react to its state changes or environment state,
- it is able to react to changes of balance between threats and protection system [8].

Security of CI objects is dynamic value. This value is measured in time and is depending on interaction of referred factors (Threat - T, Vulnerability - V), it can be of various levels. It can be expressed as:

- *sufficient security of the object;* it is when active threats cannot influence or disrupt stability and performance of the object functions,
- *condition of risk,* when object and its protecting system are able to confront active threats but this state is limit with tendency to negative change,
- *non security* when object and its protection system are not able to confront active threats, where these threats are able to disrupt performance of the functions objects even destructive changes.

From the foregoing results that the disruption of CI object security can be reach when following conditions will be fulfilled:

- there will be such changes in the object structure and attributes (in its protection system) which reduce its protecting abilities and increase its vulnerability, but the development in the environment will be without change,
- changes in environment of the object and protecting system caused threat with intensity higher then resistance.

Acknowledgments. This work has been supported by the Scientific Grant Agency of the Ministry of Education of the Slovak Republic (Projects VEGA 1/0175/14).

References

1. Zákon, č.: 454/2011 Z.z. o kritickej infraštruktúre. http://www.minv.sk/?ochrana-kritickej-infrastruktury. Accessed 29 May 2013
2. Šimák, L., a kol.: Terminologický slovník krízového riadenia.1.vyd. Žilina, FŠI ŽU, Dostupné na Internete (2005). ISBN 80-88829-75-5, http://fsi.uniza.sk/kkm/files/publikacie/tskr.pdf. Accessed 14 May 2015
3. Mikolaj, J., a kol.: Terminológia bezpečnostného manažmentu. Výkladový slovník . 1. vyd. Multiprint Košice, 191 s (2004). ISBN 80-969148-1-2
4. Lukáš, L.: Určení, rozdělení a vlastnosti detektorů narušení. In: Lukáš, L., ca kol. Bezpečnostní technologie, systémy a management I. 1. vyd. Zlín- Radim Bačuvčík VeRBuM, 316 s (2011). ISBN 978-80-87500-05-7
5. Hofreiter, L.: Ochrana kritickej infraštruktúry In: Zborník vedeckej konferencie Riešenie krízových situácií v špecifickom prostredí. ŽU, Žilina, s. 153–156 (2004)
6. Garcia, M.L.: The Design and Evaluation of Physical Protection System. Elsevier, Butterworth-Heineman, Woburn (2001). ISBN 0-7506-7367-2
7. Алаухob, С., Коцеруба, В.: Концепция безопасности и принципы создания систем физической защиты важных промышленных объектов. http://esipenko2009.narod.ru/book/KoncIntBez_1.htm. Accessed 28 May 2013
8. Hofreiter, L.: Securitológia. 1.vyd. Akadémia ozbrojených síl Liptovský Mikuláš, 138 s (2006). ISBN 978-80-8040-310-2

Different Approaches to Setting the Blast Load of Structure

Figuli Lucia[1(✉)], Jangl Štefan[1], and Picot Sylvain[2]

[1] Faculty of Security Engineering,
Univeristy of Žilina, Univerzitná 8215/1, 010 26 Žilina, Slovakia
{lucia.figuli, stefan.jangl}@fbi.uniza.sk
[2] Ecole Nationale Supérieure de Techniques Avancées de Bretagne
(ENSTA Bretagne), 2 Rue François Verny, 29200 Brest, France
sylvain.picot@ensta-bretagne.org

Abstract. The paper deals with the analysis of structure subjected to the blast load created by the explosion of ANFO explosives (ammonium nitrate and fuel oil). The paper presents the influence of blast wave idealisation on dynamical analysis of structure. Steel beam is loaded by different blast load model representing different ways of approximation of blast wave created after the explosion of explosive substances. The shape of blast pressure wave is idealised by three basic shapes – with linear decay time profile, exponential decay time profile, and exponential decay time profile considering negative phase. Dynamical analysis is done and the results from structures loaded by idealized profiles are compared with the experimental obtained results from the field test.

Keywords: ANFO explosive · Dynamic analysis · Blast wave · Numerical simulation · Critical infrastructure

1 Introduction

The easiest way how to damage or destroy various elements of critical infrastructure (mainly buildings or engineering structures) is with the use of explosives. The traditional type of explosives (as TNT, DRX etc.) are unavailable due to the restriction of European Union, so the other types of explosives, especially homemade are used. For this reason our paper is focused on the ANFO explosives.). One of such ANFO explosives which are industrially made in Slovakia is POLONIT. The explosive is a mixture of ammonium nitrate, methyl esters of higher fatty acids, vegetable oil and red dye [2]. The paper deals with the analysis of structure subjected to the blast load created by the explosion of POLONIT charge.

Structures are loaded with blast wave created as a consequence of explosion. Analysis of the structures has to be addressed to dynamic analysis because the load is changing over the time. The dynamic analysis considers various simplifications in the engineering approaches. The paper deals with various approximated models, the load is applied to the structure and the dynamic response of the model is compared with the beam response from experimental testing.

A. Kravcov et al. (eds.), *Durability of Critical Infrastructure, Monitoring and Testing*,
Lecture Notes in Mechanical Engineering, DOI 10.1007/978-981-10-3247-9_17

To study the real behaviour of blast loaded steel is fundamental for the critical infrastructure elements. The significant elements of the most important infrastructure systems in energetics, transport, ICT etc. structure are large dimension structures (high voltage pylons, pylons for the mobile communications, bridges etc.).

2 Analysis of the Structure

The paper focused on the analysis of simply supported beam subjected to the blast wave from the explosion. Standardized processes are based on idealized linear decay of blast wave profile without considering the negative pressure phase. The real shape of blast waves is different. For the analysis numerical model of structure was done and various type of load approximation was used (Fig. 1).

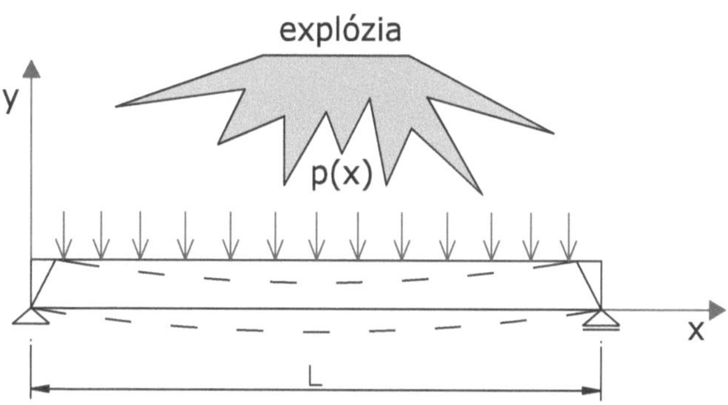

Fig. 1. Model of the blast loaded structure

The parameter of blast wave are dependent of the type of used explosives. Research on the blast loaded structures has been coming to the fore in the last years. It is evidence from the research made by [3] where the research builds on the long-time ongoing research and informs about new results of an experimental program focused on the study of blast resistance of concrete based materials and from the research made by [6] where the research is focused on the use of ultrasound method for the diagnostics of protective structures under blast load and some mathematical approaches for distance explosion was researched in [5].

All parameters, which are connected with explosions, primarily depend on the size of the explosion energy in the form of blast wave, the location of the object and its parameters and distance from the explosion [4]. The most important characteristics for blast load estimation is to know the maximum pressure. The estimation of the maximum pressure could be done using formulas described by Štoller and Rejmont [4]. Dynamical analysis of the structure has to be done. It means to find the dynamical response to the load changing in the time. Various approached can be used for example, ones described in [7–9].

3 Numerical Simulation

Model. For the numerical simulation of steel beam, VisualFEA software was used. VisualFEA is a software based on finite elements methods and it enable to do dynamic analysis of structures. The examined element is a steel beams. The beam is modelled as one dimensional element, simulated in vertical position as simply supported one.

Loading models. Diverse loading models are used for FEM structures to compare the behaviour of such loaded models and the behaviour of model with real beam. First load model is with a linear dependence of pressure on the time. It considers only positive (Fig. 2).

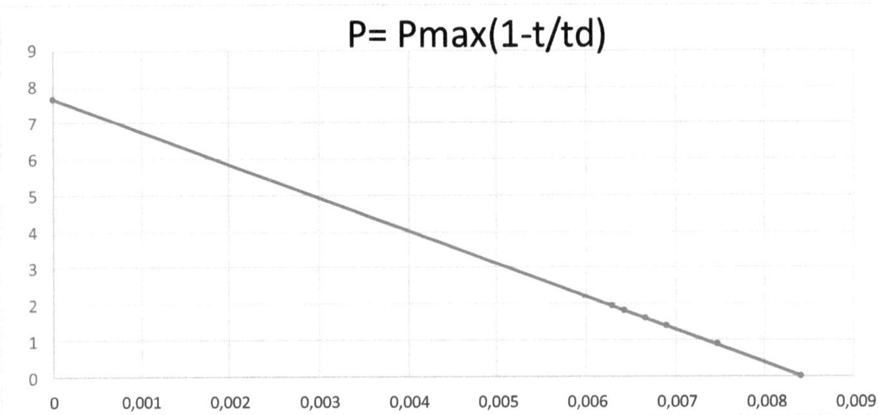

Fig. 2. Linear decay of blast wave

More complex loading model is form assumes an exponential decay (see Fig. 3) given as exponential function considering positive and negative phase of blast wave profile.

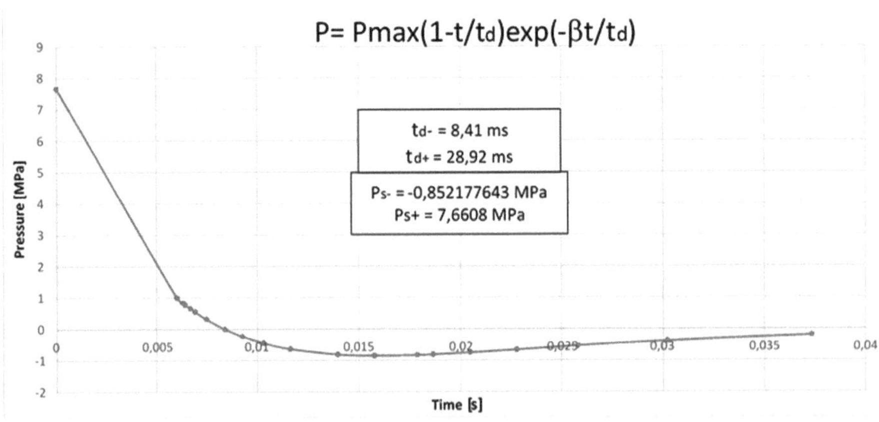

Fig. 3. Exponential decay time profile

Where t is the time after the pressure wave arrival, P_s is the peak pressure, P_0 is the ambient pressure, β is the decay parameter coefficient, t_d is the duration time. The decay parameter coefficient has been set according our previous research [1] on the value 1.1.

And as the third case study, numerical model was loaded by approximated time history diagram obtained by close approximation of real blast wave time history from the field tests (Fig. 4). The approximation was done in such way, that all characteristics points were considered.

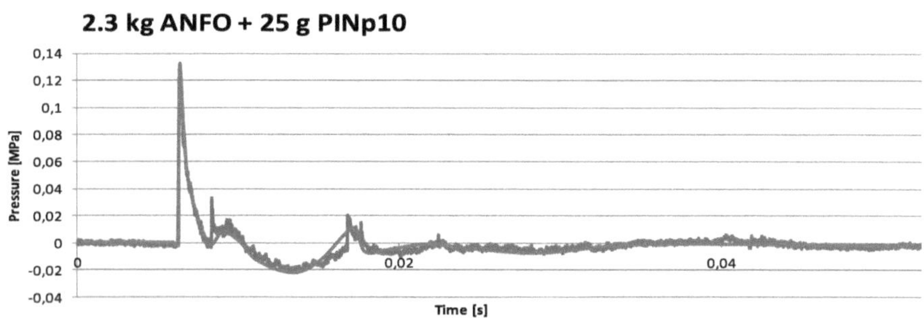

Fig. 4. Real blast wave time – history from the field tests (red) and approximation of time – history in numerical simulation (blue) (Color figure online)

4 Field Tests

The field tests were focusing on the measurement of overpressure and its influence on steel beams. POLONIT was used as explosive. The field tests took place at Military Technical and Testing Institute Zahorie. The weight of charges selected were 2.3 kg, 4.5 kg and 9 kg (the last one was not detonated because of the high overpressure of the 4.5 kg explosive which damaged the construction with steel beams). The explosives were used together with 25 g of ignition explosive PLNp10. The sensors were placed in the height of 1.6 m in the angle of 45° from the normal line in the distances 2 m, 5 m and 10 m from the source of the explosion. One of the sensors was orientated in parallel with steel beams in the distance of 5.5 m opposite of gabion wall and in the distance of 3 m from the source of explosion (we wanted to record the reflected blast wave). The explosive charges were placed at a wooden base in the height of 10 m. The structure consists of steel frame and four wide flange steel beams HEB 100 and 12 steel beams IPE 120 with span length of 1770 mm. For convenience the beams were tested in the vertical positions and simply supported. They were loaded mainly to bending caused by the blast pressure as the axial stress due to self-weight was practically negligible. All the beams were pinned at the top and roller supported at the bottom (see Fig. 5).

Fig. 5 Field tests: a) blast pressure sensors b) steel beams

Fig. 5. Field tests: (a) blast pressure sensors (b) steel beams

During the blast test pressure, strain and acceleration were measured. Maximum overpressure was measured using blast pressure sensors type 137A23 and 137A24 PCB Piezotronics. The sensors were placed in the height of 1.6 m in the angle of 45° from the normal line in the distances 2 m, 5 m and 10 m from the source of the explosion. One of the sensors was orientated in parallel with steel beams in the distance of 5.5 m opposite of gabion wall and in the distance of 3 m from the source of explosion (we wanted to record the reflected blast wave). The strain time histories were measured at the middle of beams number 2, 5, 12, 15 by surface strain sensors and accelerations by the accelerometers.

Mechanical properties are very important in determining the resistance of a structure subjected to blast loads. The mechanical properties, especially yield stress, of steel under dynamic loading condition are quite different from that under static loading. The stresses that are sustained under dynamic conditions gain values that are remarkably higher than the static compressive strength. It could be seen from the research [7, 8] (Table 1).

Table 1. Characteristics of used explosions in the field test

Beam	Static characteristics		Dynamic characteristics	
	Nominal yield	Ultimate strength	Nominal yield	Ultimate strength
HEB 100	313 MPa	441 MPa	444 MPa	625 MPa
IPE 100	330 MPa	452 MPa	468 MPa	641 MPa

The strengthening of the material S 355 during rapid deformation were taken into account. The values of static and dynamic strengths of the steel beams are in the next table.

5 Results and Comparison

The results from the numerical simulations could be compared with only small amount of the results measured in experiments. It has to be noticed that the blast pressure sensors used for measuring the blast pressure time-history in shots No 1 and 2 failed to capture any data because they were destroyed by the blast and falling debris. The same thing happened for the shot No 5 where the sensors for acceleration and strains were hit.

The dynamic analysis of the beam was done where accelerations were compared with the experimental results from field tests. Figure 7 shows the measured mid-span acceleration – time histories of the test beams and their corresponding values using the MDOF model. Generally, as can be seen in Fig. 7, model predict reasonably well both the peak and overall acceleration response of the beams.

6 Summary

The paper presents a comparison between experiments – field tests and numerical simulations for air blast loaded steel beams. It contains analysis of the dynamic response of a number of rolled I and H beams tested under the blast loads. The measured blast wave parameters were compared with models for blast load prediction. The test beams were analysed by using MDOF model based on beam finite elements. The results of models were compared with the corresponding experimental acceleration time-histories (Fig. 6).

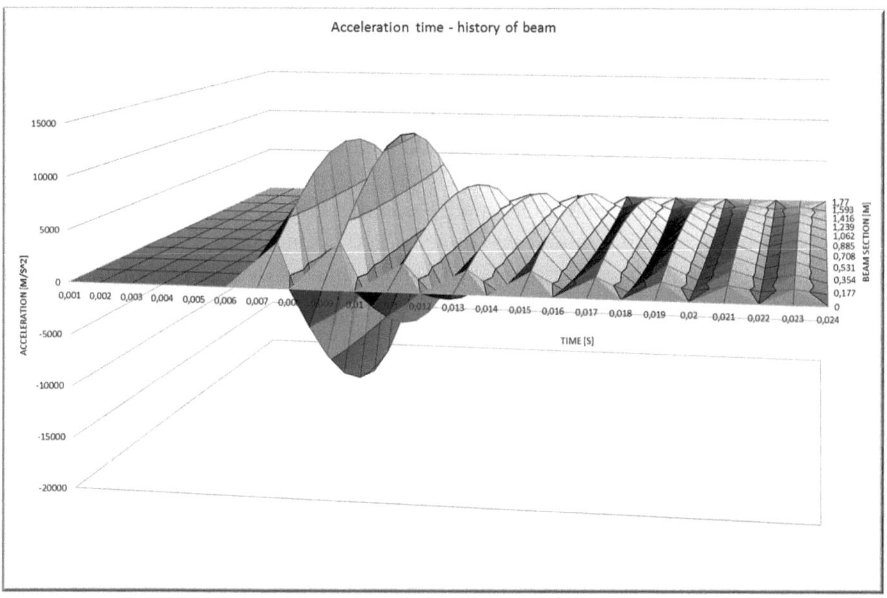

Fig. 6. Acceleration time – history of numerical model

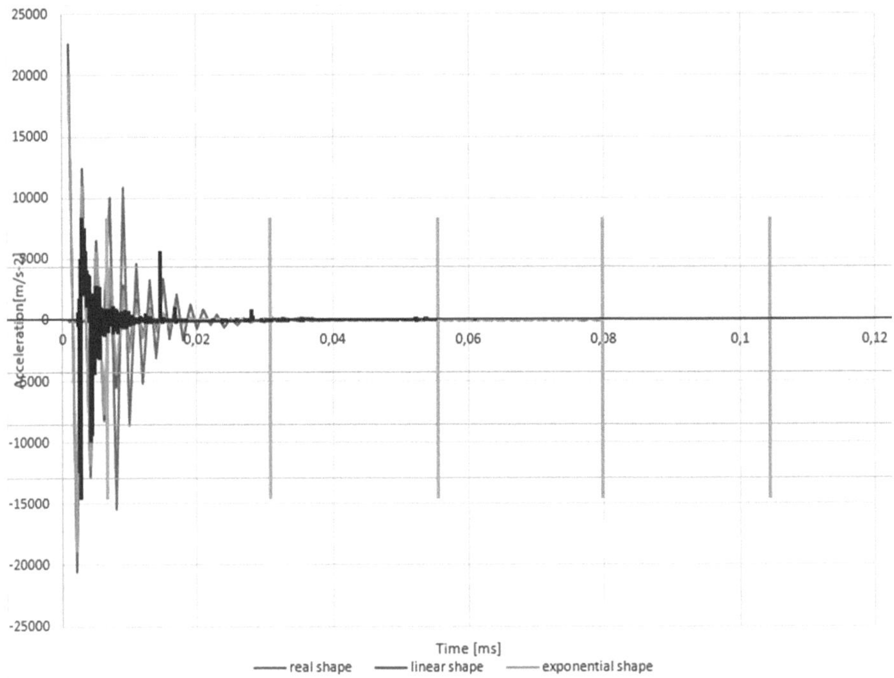

Fig. 7. Comparison of acceleration time – history of numerical model (3 difference type of loading cases) and real acceleration from the field test

For the dynamical analysis 3 type of approximation of blast wave was done. One with the linear decay, one with the exponential decay and the third one with the real time history from the experiments. The results slightly differ. For the new design or assessment of existed structures using exponential shape of load is the right one. The structure is not so overrated as using linear shape and the design is more economical.

Based on the analysis of the results and their comparison with the experimental data we could conclude that the MDOF model could predict reasonably well the blast loaded beam behaviour. It is about one dimensional beam model, and it can give very compliant results with the experimental measured ones. Advantage of such model is that it not require a large computation power and is easy to do. Analysis and publication of the new detailed numerical models are also foreseen.

Acknowledgement. This work was supported by the project VEGA 1/0240/15 Process model of critical infrastructure safety and protection in the transportation sector.

References

1. Figuli, L., Kavický, V.: Vplyv idealizovania priebehu tlakovej vlny na bezpečnosť posúdenia objektu. Riešenie krízových situácií v špecifickom prostredí (RKS 2014), zborník, Žilina (2014). ISBN 978-80-554-0872-9
2. Kavický, V.: Výbušniny v dejinách, Historická revue č (2011). ISSN 1335-6550
3. Hajek, R., Kovar, M., Foglar, M., Pachman, J., Stoller, J.: Field testing of concrete members subjected to contact and adjacent blast. Adv. Mater. Res. **1106**, 164–167 (2015). Trans Tech Publications, Switzerland
4. Štoller, J., Rejmond, J.: Design of building protection in peacekeeping operations of AČR based on simulations of the effects of blast wave. In: 5th International Conference on Military Technologies, ICMT 2015, Brno, Czech Republic, 19 May 2015 through 21 May 2015 (2015)
5. Kravtsov, A.N., Prochazka, P.P., Lok, T.S.: Assessment of laminated cylindrical arch loaded by a shock wave. Int. J. Protective Struct. **2**(2), 267–282 (2011)
6. Štoller, J., Zezulová, E.: Testing of critical infrastructure protection against the distant trinitrotoluene blast. In: Proceedings of 19th International Conference, Transport Means, Kaunas (2015)
7. Tikal, F., Špirk, S., Šedina, J.: Backstop and bullet trap design. In: 10th International DAAAM Baltic Conference on "INDUSTRIAL ENGINEERING", Tallinn, Estonia, 12–13 May 2015 (2015)
8. Papán, D., Papánová, Z.: Spectral analysis of the structures due to traffic. Procedia Eng. **91**, 171–176 (2014). ISSN 1877-7058
9. Valašková, V., Papán, D.: Theoretical frequency analysis of the historical building. Procedia Eng. **111** (2015). ISSN 1877-7058

Pressure Wave Propagation and Interaction with Structures

Zdeněk Hejmal(✉)

Faculty of Military Technology, Department of Engineer Technology, University of Defence, Kounicova 65, 662 10 Brno, Czech Republic
zdenek.hejmal@unob.cz

Abstract. The article deals with occurrence and propagation of pressure waves which can inflict substantial damage of structures and endanger the lives of soldiers and civilians. Nowadays, with the increasing incidence of terrorist acts, the use of Improvised Explosive Devices (IED) as a powerful non-conventional weapon is becoming more common. Important military objects or parts of military base entrance can be chosen as a target. For this reason, the protection of infrastructure against terrorist attacks is one of the main goals to be solved. To minimize negative effects of blast wave interaction with protective structures and shelters some knowledge of the mechanisms of explosion and shock wave propagation is necessary to provide the best available planning, design, and construction for needed disaster resistant facilities. The behaviour of the protective structure during an explosion depends entirely on the materials and design parameters used in the construction of the structure. The high performance fiber reinforced concrete (HPFRC) appears to be a promising material for protective structures. Testing of this material was carried out to assess its suitability for building civil defence shelters and pure military protective structures. The process of blast and shock wave propagation through the air and the concrete slab was numerically simulated and then suitable measuring system was designed and practical experiments were performed. The numerical simulations and the results of the experiment are compared and discussed in the article.

Keywords: Blast wave · Numerical simulation · Reinforced concrete · Seismic accelerometer

1 Introduction

Quantitative description of the blast waves can be performed in simplified manner analytically or using numerical simulation. Numerical simulation of the blast wave propagation is more time consuming but in some cases, especially when interaction with various obstacles occurs, the numerical simulation is the only alternative which provides applicable results.

In empirical approaches, air blast parameters are mostly presented for a reference explosion, and some type of scaling is subsequently used to obtain the parameter values for the actual charge weight of interest. Among the most important parameters for determining the behavior and character of the pressure wave is the distance from the point of detonation. The peak pressure, impulse and speed of the shock wave rapidly

A. Kravcov et al. (eds.), *Durability of Critical Infrastructure, Monitoring and Testing*, Lecture Notes in Mechanical Engineering, DOI 10.1007/978-981-10-3247-9_18

decreases with increasing distance but the duration of the positive phase increases. This change of parameters is most often described using Hopkinson-Cranz equation [1].

The Hopkinson scaling law states that when two charges of the same explosive material are detonated in the same atmospheric conditions, similar shock wave effects are experienced at equivalent scaled distances, Z, defined as

$$Z = \frac{R}{\sqrt[3]{W}}, \tag{1}$$

where Z is the scaled distance [m/kg$^{1/3}$], R is the real distance between a point of interest and the blast epicenter, and W is the charge weight [equivalent in kg TNT].

The simplest form of a blast wave is usually described using the Friedlander equation [2]:

$$P(t) = P_s e^{-\frac{t}{t^*}}(1 - \frac{t}{t^*}), \tag{2}$$

where Ps is the peak overpressure (70 kPa above ambient pressure) and t* is the time at which the pressure first crosses the horizontal axis (ambient pressure) before the negative phase. Thus, the value of t represents the duration of the positive phase (Fig. 1).

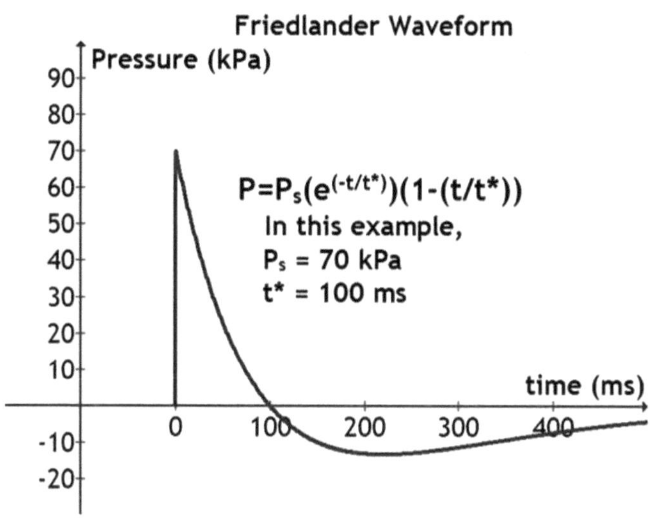

Fig. 1. The simplest form of the blast wave

The parameters describing the shock wave are experimentally determined depending on the scale parameter Z. Widely used are mainly equations derived Kinney [3]. The equation for calculating the peak pressure has the form:

$$p^o = P_a \frac{808[1 + (\frac{Z}{4.5})^2]}{\sqrt{1 + (\frac{Z}{0.048})^2}\sqrt{1 + (\frac{Z}{0.32})^2}\sqrt{1 + (\frac{Z}{1.35})^2}}, \quad (3)$$

where p^o is the overpressure (the difference between peak pressure and ambient pressure), the P_a is the ambient atmospheric pressure and the item Z is a scaled distance, Eq. (1).

2 Numerical Simulation

There are many programs which can be used for simulation of the explosion and shock wave propagation nowadays. The most commonly used include LS-DYNA, AUTODYN, ABAQUS, CONWEP, Air3D and others. Due to the availability and relatively easy control the software ANSYS AUTODYN was used. The main purpose of this numerical simulation was to compare numerical and analytical approach. The spherical blast and the propagation of the blast wave was performed in 2D axial symmetric model and standard library materials with following parameters were used:

1. Air

- reference density 1.225 kgm^3,
- equation of state "Ideal Gas" (only the value of adiabatic exponent 1.4 needs to be supplied),

2. TNT

- reference density 1630 kgm^3,
- equation of state Jones-Wilkins-Lee,
- parameter A = 3.7377 kPa,
- parameter B = 3.7471 kPa,
- parameter R1 = 4.15,
- parameter R2 = 0.9,
- parameter W = 0.35,
- C-J detonation velocity 6930 ms^{-1},
- C-J energy per unit volume 6 GJm^{-3},
- C-J pressure 21 GPa.

3. Solver: Euler, 2D Multi-material.

The air space of the wave propagation was modelled by wedge geometry with appropriate grade zoning. Flow_out boundary was used to simulate open space in the direction of the blast wave propagation. The gauges were placed equidistantly with spacing of 10 m. The simulation of the blast wave propagation using 2D axial symmetric model is in Fig. 2.

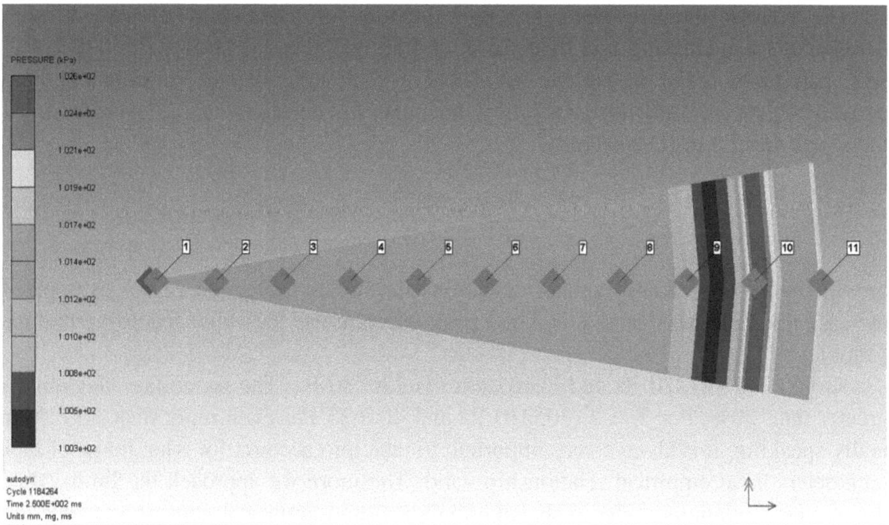

Fig. 2. The simulation of the blast wave propagation using the 2D axial symmetric model

The data from the AUTODYN environment were exported as a comma separated files and imported into MATLAB computing system for further data processing.

To compare the results of the analytical and numerical approach the same problem was solved using Kinney equation for calculating the peak pressure (3). The results for both methods are shown in Fig. 3.

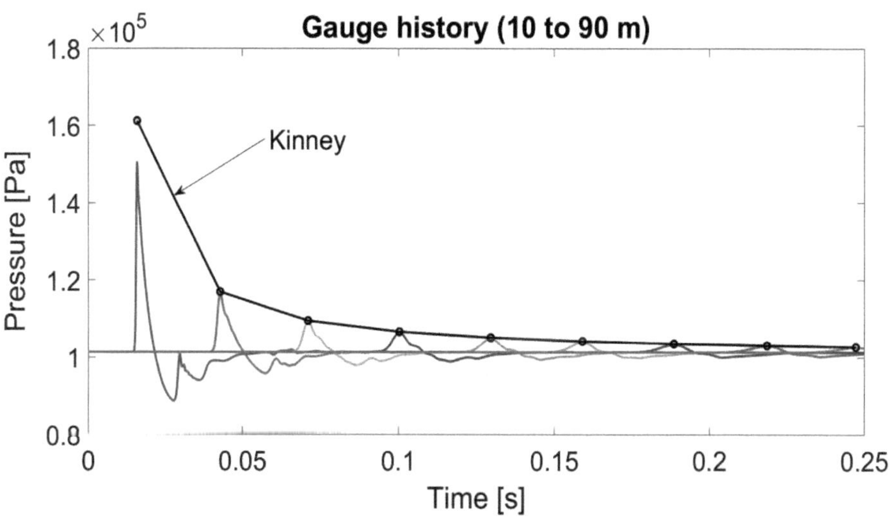

Fig. 3. The comparison of the two calculation methods for propagating of the pressure wave

The results obtained using the two methods of calculation showed excellent agreement for increasing real distance R and accordingly increasing the value of the scale parameter Z. For the parameter values Z > 5 the difference in values is less than 2 percent. Other recommended analytical formulas for determining the overpressure in the shock front e.g. (Denkstein)

$$\Delta p_+ = \frac{93}{Z} + \frac{383}{Z^2} + \frac{1275}{Z^3}, \quad [2 \leq Z \leq 200] \tag{4}$$

provide even substantially greater deviations for smaller values of Z. For example, in the case the scale parameter Z = 3, the peak pressure has the value according the used formula:

Kinney (3) 181270 Pa and Denkstein (4) 220780 Pa. The percentage deviation is greater than 20%. For Z = 2 (305210 Pa and 401625 Pa) even more than 30%. Generally speaking, it is always very important to take into account for what range of input parameters is an empirical relationship valid. The more we approach the limit values, the greater deviation can be expected. Analytical and empirical approach is fully justified for solving pressure waves propagation without considering the interaction with the external environment and objects. It is almost always recommended to use numerical simulations if it is necessary to take into account the effects of interaction and interference.

3 Blast Field Tests

There are several effects caused by the explosion of the charge. In the air above the ground or directly on the surface of any object. The explosion of the charge is classified as a surface blast when the charge explodes on the ground or in the maximum height of 20 m above the ground. As a result of the blast some amount of energy causes seismic shock and propagation of stress waves through the ground. The primary goal of these tests was to determine the behaviour, resistance and response of high performance reinforced concrete which was exposed to blast loading. The secondary objective was to detect the intensity of seismic effects under given conditions. The tests were carried out in the military training area Boletice. The explosive charges, their setting up and ignitions were provided by 15th Engineer Regiment.

The tests were performed in cooperation with Czech Technical University in Prague, Faculty of Civil Engineering and University of Pardubice, Institute of Energetic Materials. The real scale fibre reinforced concrete slabs with dimensions 6 × 1.5 × 0.3 m were under tests. The material of the slabs was varying in fibre content, fibre type and strength, concrete matrix strength and surface coating. The slabs were placed on wooden logs and the soil under the slabs was excavated. These samples were loaded by blast of explosive charge 25 kg TNT placed 0.5 m above the slab. The process of the blast loading and the blast shock wave propagation were recorded using high speed camera. The setup of the test is shown in Fig. 4. The slab after performed explosion and before the visual inspection and subsequent ultrasonic non-destructive test is shown in Fig. 5.

Fig. 4. The setup of the experiment (slab 6 × 1.5 × 0.3 m)

Fig. 5. The slab after explosion and before ultrasonic non-destructive test

In parallel with this experiment smaller concrete slabs were tested. The test samples had dimensions 0.1 × 1.00 × 1.00 m and 0.15 × 1.00 × 1.00 m. The slabs were placed in a rack made of thin-walled closed steel profiles. The racks were fixed against displacement using wooden stakes. The charges were situated in the selected locations (contact blast) and their weight ranged from 300 to 500 g. Military explosive PlNp 10 was used during the experiment. This is a plastic explosive with high efficiency and detonation velocity 7600 ms^{-1}.

Both tests supplied valuable information about the behaviour of the tested materials under extreme load conditions. This information is essential for the appropriate design and technology of building protective structures and facilities. The obtained information helps also to improve further the quality of these materials and also to create a credible material models for meaningful numerical simulation. It is the only way to reduce the extremely high overall costs of the real tests using explosives.

4 Measurement of Seismic Effects

To obtain information about the intensity of seismic effects caused by the realization of tests using explosives it was necessary to propose suitable measuring chain and to determine the appropriate measurement points. Based on the analysis of current knowledge, we decided to measure the vibration of the ground using seismic accelerometers placed radially to the detonation point. The goal was also to determine the extent to which the vibrations of the ground is propagating through the soil and how much energy is transferred by acoustic air pressure waves. We have used 3 sensors placed inline at defined distances from the centre of the blast. The first sensor was in the distance of 265 m, the second in 300 m and the third one in 330 m. The setup is shown in Fig. 6.

The red circle in Fig. 6 represents the detonation point and the blue circles are the placements of individual sensors. The attachment of transducers was performed by special steel pins that were sunk into the ground to faithfully transfer the movement of the ground and relative motion of sensor to the ground was completely eliminated. The attachment of accelerometers KB12VD to the ground by steel pins mentioned above is shown in Fig. 7.

For the acceleration measurement the high sensitive seismic accelerometers KB12VD were used. These accelerometers are suited for seismic measurement and building vibration, particularly at low frequencies. Extremely sensitive piezoelectric system provides excellent resolution and the lowest possible noise. Basic parameters are as follows:

- IEPE output,
- Voltage sensitivity 1000 ± 10% $mVm^{-1}s^2$,
- Range ±5.5 ms^{-2},
- Destruction limit 2000 ms^{-2},
- Linear frequency range $f_{5\%}$ 0.25 ÷ 130 Hz.

The signals from accelerometers were filtered, amplified and sampled by DEWE-50-USB2-8. It is an eight-channel measuring device with USB interface that

Fig. 6. The placement of seismic accelerometers for the detection of seismic effects (Color figure online)

Fig. 7. The attachment of seismic accelerometers KB12 V, on the left side arrangement for vertical direction and on the right for horizontal direction acceleration measurement

enables simultaneous sampling of eight channels. Depending on the type of the measured quantity and type of sensor it is necessary to use appropriate measuring module, in our case, we used DAQP-ACC-A, which is a measurement module with IEPE input and with the following basic parameters:

- Ranges ±5 V, ±1.66 V, ±500 mV, ±166 mV, ±50 mV,
- Gain 1, 3, 10, 30, 100, gain error max. 0.5%,
- Sensor excitation 4 or 8 mA,
- Highpass filters 0.5 Hz or 5 Hz,
- Lowpass filters 1 kHz, 10 kHz, 100 kHz, 300 kHz,
- Typical SNR for Gain = 10 equals 91 dB,
- Output voltage ±5 V.

The results of vibration measurements of soil in a horizontal plane perpendicular to the direction of the explosion are illustrated in Fig. 8.

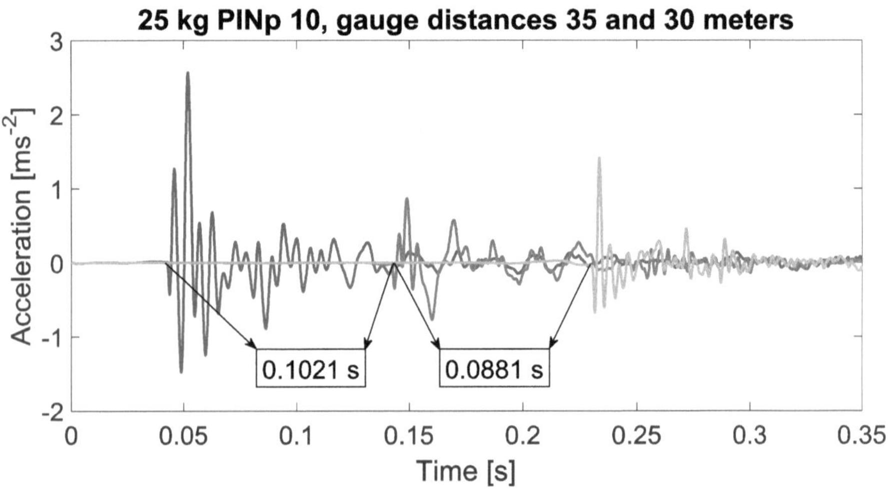

Fig. 8. The results of acceleration measurement

The maximum value of the measured acceleration of the ground caused by detonation of 25 kg of plastic explosive PlNp10 at a distance of about 250 m is up to 3 ms^{-1}. From the signal delays of individual sensors and their distances, the velocity of the seismic excitation can be calculated. The detected velocity slightly above 340 ms^{-1} indicates that most of the energy is spreading further into space through air and not through the soil or subsoil.

5 Summary

The issue and, if possible, the real ability to determine the parameters and the effects of pressure waves generated by the explosion is for designing of protective structures extremely important.

Analytical and empirical methods for solving this problem are fast and in many cases provide satisfactory results. In the case of solving problems where you need to take into account the complex interactions and interference the numerical simulation is often the only possible solution to the problem. The credibility of numerical results, however, will be always influenced by a considerable number of parameters of the computational model, and last but not least by adequate description of the material properties. This poses a particular problem for the numerical simulation of new materials and not enough information is available.

The measurement of seismic disturbances caused by surface blast indicates that the primary danger for the damage of the ground structures represents the pressure wave propagation. Induced ground vibration plays from this point of view only marginal role. The work presented in this article has been supported by the Czech Republic Ministry of Defence - K 201 Department development program "Development of technologies in weapons construction, ammunition, weapons instrumentation, materials engineering and military infrastructure".

References

1. Guzas, E.L., Earls, C.J.: Air blast load generation for simulation structural response. Steel Compos. Struct. **10**(5), 429–455 (2010). doi:10.12989/scs.2010.10.5.429. Accessed 18 Feb 2016
2. Friedlander, F.G.: The diffraction of sound pulses. I. Diffraction by a semi-infinite plate. Proc. R. Soc. A Math. Phys. Eng. Sci. **186**(1006), 322–344 (1946)
3. Kinney, G.F., Graham, K.J.: Explosive Shocks in Air, 2.vyd. Springer, New York (1985). ISBN 978-3-642-86684-5
4. Štoller, J., Maňas, P., Zezulová, E.: Blast testing and simulation methods. In: Praha: Česká Technika nakladatelství ČVUT - Praha. Blast Testing and Simulation Methods. Neuveden, pp. 35–100 (2015). ISBN 978-80-01-05898-5
5. Štoller, J., Dvořák, P.: Field tests of high performance fiber reinforced concrete slabs: impact of contact and distant explosions. In: ICMT 2015 - International Conference on Military Technologies 2015, art. no. 7153679 (2015). ISBN: 978-807231976-3, doi:10.1109/MILTECHS.2015.7153679
6. Štoller, J., Dvořák, P.: Cementitious composite materials for critical infrastructure. In: Transport Means 2016 - Proceedings of 20th International Scientific Conference, Kaunas University of Technology, Kaunas, Litva, pp. 9–10 (2016). ISSN 1822-296X
7. Štoller, J., Dvořák, P.: Non-destructive testing of full-scale slabs before and after blast load. In: Transport Means - Proceedings of the International Conference, January 2015, pp. 298–301 (2015). ISSN: 1822296X

Laser-Ultrasound Imaging for the Investigation of Heterogeneous Media

Alexander A. Karabutov(✉), Elena B. Cherepetskaya, Anton S. Bychkov, and Nikolay A. Morozov

Moscow Mining Institute, The National University of Science and Technology MISiS, Leninski Prospect 4, Moscow 119991, Russia
{aak, abychkov}@optoacoustic.ru, echerepetskaya@mail.ru, morozov499@yandex.ru

Abstract. The development of a laser ultrasonic imaging technique for diagnostics of heterogeneous materials involves solving the inverse problem of ultrasound scattering. The paper describes the back-projection algorithm for the reconstruction of laser-ultrasonic images of inhomogeneous media. To demonstrate the capabilities of laser ultrasonic imaging, an experiment was carried out to reconstruct the image of an orifice 1 mm in diameter in a duralumin half-cylinder with a radius of 65 mm.

Keywords: Laser-ultrasound imaging · Ultrasound scattering · Back-projection algorithm

1 Introduction

The wide use of composite materials, multilayer composites, and geopolymers in various structures subjected to sustained axial and transverse loads calls for mathematical models describing micro and macro destruction of these materials and new experimental techniques and systems for the diagnostics of their structure and properties.

Most widely used for the diagnostics of heterogeneous media are X-ray and ultrasonic [1] imaging techniques. Laser-ultrasonic (LU) structuroscopy showed to be prominent to solve this task [2]. This technique involves the excitation of elastic waves by a laser pulse absorbed in a thin surface layer of the test object or a dedicated optoacoustic (OA) generator, the propagation of the ultrasonic pulse in the test object, and the registration of transmitted or scattered ultrasound waves with high temporal resolution. The shape and amplitude of the probe pulse depend on the characteristics of the OA generator and the laser pulse. The probe pulses have a smooth unipolar temporal profile free of reverberations. The spectrum of these pulses is wide, limited only by the spectral width of the laser pulse envelope. Laser ultrasonic techniques are most used in non-destructive testing, studying the structure of materials and industrial products and biological objects diagnostics [3–8].

The longitudinal spatial resolution is high due to the short duration of the probe pulse; the signal bandwidth is appropriate, though. The small diameter of the probe beam provides enhanced sensitivity. The aperiodicity of the probe pulse provides

A. Kravcov et al. (eds.), *Durability of Critical Infrastructure, Monitoring and Testing*, Lecture Notes in Mechanical Engineering, DOI 10.1007/978-981-10-3247-9_19

almost the absence of "dead zone". This becomes possible due to ultrasonic transducer operation as a receiver only.

Most studies in acoustic imaging currently focus on the enhancement of transceivers and the improvement of algorithms of image reconstruction [9] and signal processing, for example, with the help of wavelet analysis [10].

Present study focuses on the development of a laser-ultrasound imaging technique using a multi-element detector array to investigate the structure and properties of heterogeneous media.

2 Back-Projection Algorithm for the Reconstruction of Laser-Ultrasonic Images

The development of a laser ultrasonic diagnostic technique involves solving the inverse ultrasound scattering problem, which consists of determining the distribution of density and compressibility in the test object using pressure signals $P(\vec{r}, t)$, which, ideally, are supposed to be measured at every point of certain surface S. The direct ultrasound scattering problem can be described by the following equation [11, 12]:

$$\nabla^2 P(\vec{r}, t) - \frac{1}{c^2}\frac{\partial^2 P(\vec{r}, t)}{\partial t^2} = \frac{1}{c^2}\frac{\partial^2 P(\vec{r}, t)}{\partial t^2}\tilde{\beta}(\vec{r}) + \operatorname{div}[\tilde{\rho}(\vec{r}) \cdot \operatorname{grad} P(\vec{r}, t)], \qquad (1)$$

where $\tilde{\beta}(\vec{r}) = (\beta(\vec{r})/\beta_0 - 1)$ is the relative compressibility fluctuation and $\tilde{\rho}(\vec{r}) = (1 - \rho_0/\rho(\vec{r}))$ is the relative density fluctuation, $c = (\rho_0\beta_0)^{-1/2}$ is the sound velocity in homogeneous medium. In this paper, the problem is addressed using the Born and acoustic far-field approximations for the case of weak scatterers.

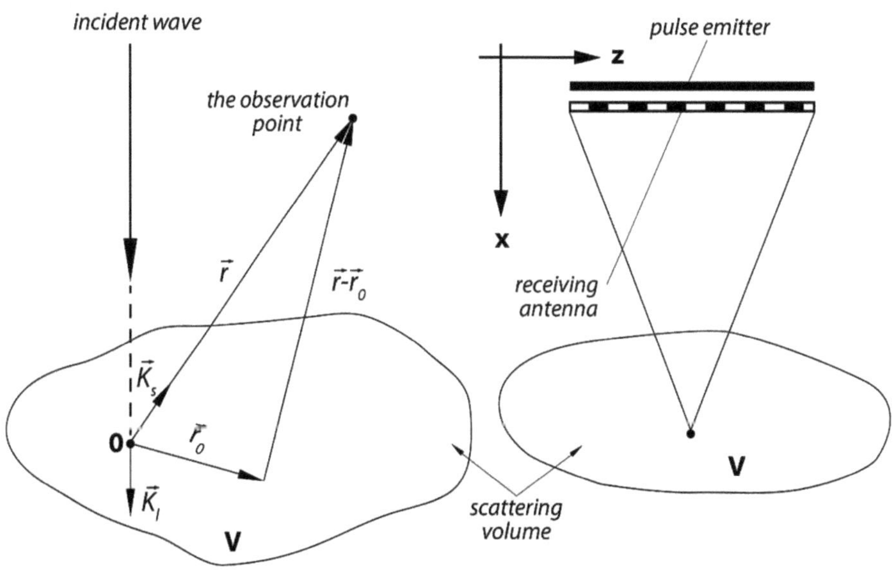

Fig. 1. Diagram of scattering by a single obstacle of arbitrary shape.

When the incident wave is a plane wave with amplitude P_0 and angular frequency ω, the following solution can be expressed as (the schematic is depicted in Fig. 1) [12]:

$$P_S(\vec{r},t) = P_0 \frac{e^{1\left(\omega t - \vec{k}_l \vec{r}\right)}}{4\pi|\vec{r}|} \iiint_V k_l^2 \left(\tilde{\beta}(\vec{r}_0) + \tilde{\rho}(\vec{r}_0)\cos\theta\right) e^{1\vec{K}\vec{r}_0} d^3\vec{r}_0, \tag{2}$$

where $\left|\vec{K}\right| = \left|\vec{k}_l - \vec{k}_s\right| = 2k\,\sin\theta/2$ is the phase delay between the signals scattered by different heterogeneities; θ is the angle between the wave vectors of the incident and scattered acoustic waves, $\vec{k}_l = \vec{k}$ is the wave vector of the incident wave.

Equation 2 is similar to the Rayleigh integral in optoacousic tomography using wave fields rescattered by point sources inside the scattering volume V.

When the temporal profile of the OA probe pulse is known, Eq. 2 is integrated over frequency. Factor k_l^2 corresponds to the second derivative of the incident OA pulse with respect to time. In view of the above, we can write:

$$P_S(\vec{r},t) = \frac{P_0}{4\pi|\vec{r}|} \iiint_V \left(\tilde{\beta}(\vec{r}_0) + \tilde{\rho}(\vec{r}_0)\cos\theta\right) \frac{1}{c^2} \frac{\partial^2 f_0(\tau)}{\partial \tau^2} d^3\vec{r}_0, \tag{3}$$

where τ is the retarded time, $f_0(\tau)$ is the normalized temporal profile of the probe pulse. The inverse problem of the scattering of broadband OA signals can be regarded as the problem of the reconstruction of some function of sources $\varepsilon(\vec{r}_{sc})$, which is responsible for density and compressibility fluctuations in the medium under investigation, by using backscattered OA signals registered by the receiver array.

Mathematically, this means the determination of function $\varepsilon(\vec{r}_{sc})$ from the inhomogeneous wave equation:

$$\frac{\partial^2 p}{\partial t^2} - c_0^2 \nabla^2 p = p_{sc}(\vec{r}_{sc}, \tau), \tag{4}$$

$$p_{sc}(\vec{r}_{sc}, \tau) = p_{sc}(\vec{r}_{sc}) \frac{\partial^2 f_0(\tau)}{\partial \tau^2}, \tag{5}$$

$$p_{sc}(\vec{r}_{sc}) \approx p_0 \varepsilon(\vec{r}_{sc}), \tag{6}$$

where $= t - x_{sc}/c$, $\vec{r}_{sc}$ is the radius vector of scatterer, and x_{sc} is the coordinate of the scatterer along the direction of the incident wave (Fig. 1).

Moreover, the ultrasound scattering inverse problem in LU imaging can be solved using the back-projection method just like the inverse problem in OA tomography. In 2D LU imaging, the image is likewise produced by projection of "probability arcs" from the points corresponding to the positions of individual receiving elements. Here we show that in the case of LU imaging the "probability arcs" are parabolas (see Fig. 2) with the vertices above the receiving elements and the branches extending in the direction opposite to the x axis.

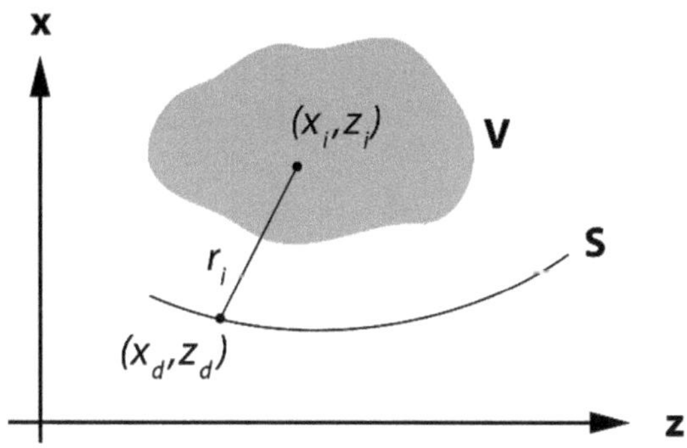

Fig. 2. To the calculation of "probability arcs" in the case of 2D LU imaging.

In the case of 2D imaging, distance R that the ultrasonic signal travels from the transmitter coinciding with the z axis (Fig. 2) to one of points (x_i, z_i) of scattering volume V and from that point to the receiver located at point (x_d, z_d) on the surface S: $R = x_i + r_i = ct_i$, where $r_i^2 = (x_i - x_d)^2 + (z_i - z_d)^2$ is the square of the distance from a point in the scattering volume to the receiving element and t_i is the signal propagation time. Then the dependence of x_i on z_i at constant t_i, which is the equation for probability arcs, is as follows:

$$x_i = \frac{1}{2}\frac{x_d^2 + (z_i - z_d)^2 - c^2 t_i^2}{x_d - ct_i}. \tag{7}$$

at $t_i > x_d/c$. The condition for t_i is a consequence of the causality condition because the signal from a scatterer cannot reach the receiver before the probe signal that propagates from the transmitter in straight line.

3 Experiment

Here we describe an experiment to demonstrate the capabilities of the laser ultrasonic imaging technique. An orifice with diameter $d = 1$ mm was drilled in a duralumin half-cylinder with radius $R = 65$ mm, at distance $R - L = 25$ mm from the axis of the half-cylinder (see Fig. 3). Emitted by a pulsed Nd:YAG laser, the laser light with a wavelength of 1064 nm was absorbed by a special polymer layer acting as an optoacoustic generator. Laser energy absorption was followed by the excitation of a short ultrasonic probe pulse. The probe pulse propagated through the test object and was scattered at the orifice. Scattered signals were recorded by one piezoelectric element whose position changed from $\alpha = 15°$ to $\alpha = 70°$ and from $\alpha = -15°$ to $\alpha = -70°$ with step $\Delta\alpha = 5°$. Thus, we simulated a receiver array consisting of 24 piezoelectric elements. Then, the recorded ultrasonic signals were processed using the back-projection

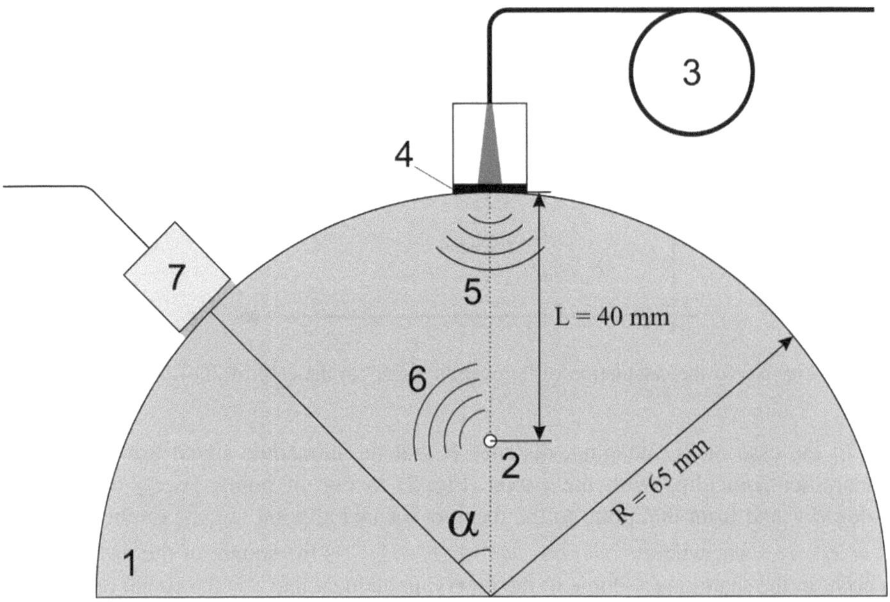

Fig. 3. Schematic of experiment with 65 mm-radius duralumin half-cylinder (1) with orifice 1 mm in diameter (2): Laser light is delivered through optical fiber (3) and is absorbed by optoacoustic generator (4) where a short ultrasonic pulse (5) is formed and then scattered by the orifice. Scattered pulse (6) is registered by receiving piezoelectric element (7) whose position (angle α) varies.

algorithm to locate the position of the orifice. A resulting map of the spatial distribution of acoustic scatterers in the test object is presented in Fig. 4. The orifice is clearly seen. Note that the orifice is imaged in blue since the probe pulse was reflected at the free surface out of phase. Archwise artifacts are also clearly seen; such artifacts are characteristic for the back-projection algorithm and associated with the finite number of receiving elements.

The use of an array of receiving piezoelectric transducers will make it possible to examine rock specimens and building materials in real time. 2D laser ultrasonic (LU) imaging has higher spatial resolution than 1D LU structuroscopy because a map of the distribution of scatterers is based on the recordings of waves scattered at different angles. The spatial resolution of LU tomography depends on the frequency band of the receivers, the velocity of sound waves, the size of the receiving elements, and the geometry and aperture of the receiver array.

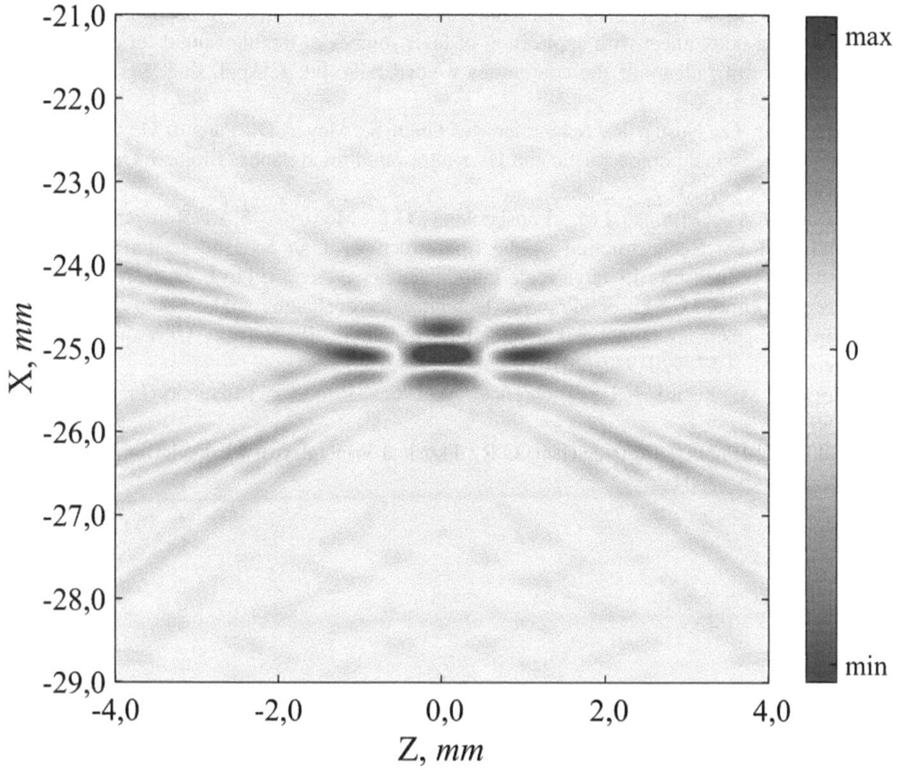

Fig. 4. Map of the spatial distribution of scatterers in duralumin half-cylinder across an area 8 mm × 8 mm in size near the orifice. (Color figure online)

Acknowledgements. This work was carried out with financial support from the Russian Science Foundation (grant no. 16-17-10181).

References

1. Gholizadeh, S.: A review of non-destructive testing methods of composite materials. Procedia Struct. Integrity **1**, 50–57 (2016)
2. Gusev, V.E., Karabutov, A.A.: Laser Optoacoustics. American Institute of Physics, New York (1993)
3. Kozhushko, V.V., Hess, P.: Laser-induced focused ultrasound for nondestructive testing and evaluation. J. Appl. Phys. **103**, 124902 (2008)
4. Bernstein, J.R., Spicer, J.B.: Hybrid laser/broadband EMAT ultrasonic system for characterizing cracks in metals. J. Acoust. Soc. Am. **111**, 1685–1691 (2002)
5. Simonova, V.A., Savateeva, E.V., Karabutov, A.A.: Novel combined optoacoustic and laser-ultrasound transducer array system. Mosc. Univ. Phys. Bull. **64**, 394–396 (2009)
6. Zhang, X.R., Zhang, W., Wang, X.D., Zhang, L.D.: Laser ultrasound characterization of chemically prepared nano-structured silver. Appl. Phys. A **70**, 573–580 (2000)

7. Suslov, O.A., Novikov, A.A., Karabutov, A.A., Podymova, N.B., Simonova, V.A.: Use of acoustoelasticity effect with application of laser sources of the ultrasound for control of the stress state of railbars of the continuous welded rails. Int. J. Appl. Eng. Res. **10**, 41121–41128 (2015)
8. Wurzinger, G., Nuster, R., Schmitner, N., Gratt, S., Meyer, D., Paltauf, G.: Simultaneous three-dimensional photoacoustic and laser-ultrasound tomography. Biomed. Opt. Express **4**, 1380–1389 (2013)
9. Burov, V.A., Grishina, I.M., Lapshenkina, O.I., Morozov, S.A., Rumyantseva, O.D., Sukhov, E.G.: Reconstruction of the fine structure of an acoustic scatterer against the distorting influence of its large-scale inhomogeneities. Acoust. Phys. **49**, 627–637 (2003)
10. Yang, J., Ume, I.C.: Laser ultrasonic technique for evaluating solder bump defects in flip chip packages using modal and signal analysis methods. IEEE Trans. Ultrason. Ferroelectr. Freq. Control **57**, 920–932 (2010)
11. Morse, P.M., Ingard, K.U.: Theoretical Acoustics. Princeton University Press, Princeton (1987)
12. Hill, C.R., Bamber, J.C., ter Haar, G.R.: Physical Principles of Medical Ultrasonics. John, West Sussex (2004)

Designing Military Constructions During Crisis Situation and New Elevation Data of the Czech Republic

Jan Sobotka(✉) and Klára Cibulová

University of Defence, Kounicova 65, 662 10 Brno, Czech Republic
{jan.sobotka,klara.cibulova}@unob.cz

Abstract. A really important measure of ensuring of the defense of the Republic is the Operational Preparation of National Territory. The Operational Preparation of National Territory also includes management of crisis situations especially during the attack critical infrastructure. The paper is focused on the possibilities for designing of temporary bridges and temporary roads in the destruction or disruption of the Critical Infrastructure. This paper shows the examples to use a new elevation data from airborne laser scanning country for the designing of military buildings. These data may serve as a base for creating digital terrain model of an area of interest. Subsequently, this digital terrain model is the base for designing of temporary bridges or temporary roads.

Keywords: Critical infrastructure · Temporary bridge · Temporary road · Elevation data · Designing · Digital terrain model

1 Introduction

An effective system for ensuring of the defense of the Czech Republic (CR) consists of a combination of policy measures the highest levels of government (legislative and executive), structure of the Czech Army, the measures of the Operational Preparation of National Territory, also the National Action Response System for crisis management needs of the National Defense, selectively adding Armed Forces of the Czech Republic, a mobilization of Army forces CR and other measures. This system is closely departmentally and interdepartmentally linked, the individual system elements constitute the basic prerequisites for the development of other elements and they are dependent on each other.

One of the key measures is the Operational Preparation of National Territory (OPNT). OPNT is summary of military, economic and defensive character, which are planned and carried out in peacetime under state of emergency or state of war. The main task is to build necessary conditions for the fulfillment tasks of the armed forces and security needs of the population on the territory of the Czech Republic [1].

The Critical Infrastructure consists of components or systems elements (buildings, equipment, facilities or public infrastructure) and their operators. A functional impairment would have a serious impact on national security, a securing of the basic living needs of the population, health or economy of the state [2]. Critical infrastructure

A. Kravcov et al. (eds.), *Durability of Critical Infrastructure, Monitoring and Testing*, Lecture Notes in Mechanical Engineering, DOI 10.1007/978-981-10-3247-9_20

elements define ministries and central government authorities. Their operators process the *Plan of emergency preparedness subject of critical infrastructure* to protect and safeguard actions during crisis situations.

Every state, including Czech Republic, needs an infrastructure that will ensure the fulfillment of their basic functions and thus the proper functioning of the state. The priority is to ensure military stability, sovereignty, territorial integrity and democratic foundations of the state. In times of peace to the fore the threat of natural disasters, technical and technological accidents and the associated need to protect lives and health of people, property and environmental protection, security, internal security and public order. To achieve these objectives, it is necessary to have a functioning infrastructure, which is essential, vital - critical. To ensure the protection of the Critical Infrastructure, it is necessary to know the threats that could cause malfunction and thereby jeopardizing the already-mentioned basic functions of the state. If we know these threats, we can effectively devise measures to minimize the risk of destruction or attack critical infrastructure, or the possibility of a quick recovery. The paper is focused only on the area of critical transportation infrastructure, particularly on the new possibilities (utilization of the new elevation data of the CR) for designing of the temporary bridges and temporary roads. Destruction or disruption of the bridge on the Critical Infrastructure is a typical example of possible breach it´s functionality. And it does not matter whether it is a military or non-military crisis situation.

2 Joint Exercise of Geodetic Service and Engineers

Geodetic Service and Engineers have closely collaborated together till 2002, when catastrophic floods destroyed a large number of bridges all over the country. Geodesists hat to survey many places, where bridges were damaged. These geodetic measurements in situ were the basis for temporary bridge designing. This cooperation continued in following years, especially due to the fact that the floods destroyed bridges and road infrastructure on regular annual intervals. Joint exercises are regularly held since 2006 and the main goal is to keep good trained geodesists and engineers in dealing with crisis situations.

The joint exercise of Geodetic Service and Engineers is attended by more than 50 soldiers this year (especially from Military Geographic and Hydrometeorological in Dobruška and engineers from the University of Defence in Brno). The main task of this exercise was to verify declared capabilities of Geodetic Service and Engineers to solve the crisis situation (especially in the area of designing and construction of temporary bridges). The theme of the exercise was destroyed bridge on a first class road across the river Labe in Litoměřice (object of the Critical Infrastructure). The main goal of this exercise was to design the project of temporary bridge (including temporary roads and handling area near the bridge) on the basis of available geodetic data. A geodetic measurement in situ was a basis for temporary bridge project and digital terrain model of 5th generations was a basis for temporary roads and handling area near the bridge.

3 Digital Elevation Database of the Czech Republic Territory

These new elevation data are the result of the airborne laser scanning of the Czech Republic (2009–2013). The new database will replace the old one, which is obsolete and inoperable. The old elevation database is represented primarily by Digital Terrain Model of the 3rd generation (DTM3G). Figure 1 shows DTM 3G in a Triangulated Irregular Network – TIN, which was created in the special drawing software AutoCAD Civil 3D. The Ministry of Defense of the Czech Republic created this model using stereo photogrammetry in 2003 – 2008. It has a mean square error of an elevation 1 – 2 m in the open areas, 1 – 2 m in the urban zones and 3 – 7 m in the forest. DTM 3G is provided in the GRID form by segment 10 × 10 km. Altitudes of the terrain are determined for nodal points of network 20 × 20 m [3].

Digital terrain model of 4th and 5th generation (DTM4G ad DTM5G) come from Airborne Laser Scanning of the Czech Republic. Airborne Laser Scanning uses the method of Light Detection and Ranging (LIDAR). LIDAR for terrain and land surveying has made significant contributions to many environmental, engineering and civil applications and it has been used increasingly by the public and commercial sectors since the early 1990s [4]. An airborne LIDAR data acquisition system estimates the distance between the instrument and a point on the surface by measuring the time the laser pulse needs to return. Differential Global Positioning System (DGPS) and Inertial Navigation System (INS) complement the data with position and orientation, respectively. The laser's capability to penetrate gaps in vegetation allows measuring occluded ground, as at least first echo (FE) and last echo (LE) can be recorded. Hence, LIDAR is superior to traditional photogrammetric techniques such as stereo matching [5].

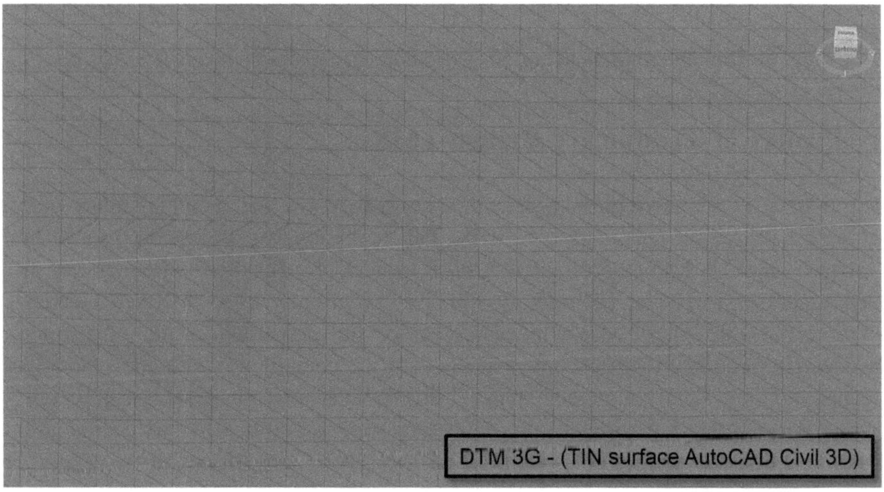

Fig. 1. DTM3G (size of the displayed area is 500 × 500 m).

Furthermore, as LIDAR is an active sensor, it is independent from external light conditions, and is therefore not affected by shadows and can be used at night [6]. DTM4G is provided in the GRID form (5 × 5 m) with a mean square error of an elevation of 0.3 m in the open areas and 1 m in the forest. The whole model of the Czech Republic will be created within six months after scanning in 2012 [7]. DTM5G (Fig. 2) is provided in the TIN form with a mean square error of an elevation of 0.18 m in the open areas and 0.3 m in the forest. The whole model of the Czech Republic will be created within two years after scanning in 2015 [8]. More about accuracy of the new generation models see in [9–11].

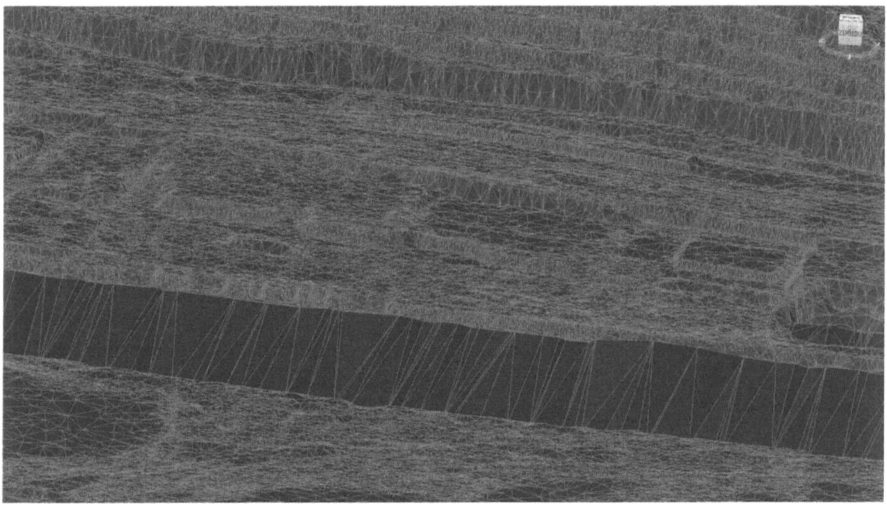

Fig. 2. Digital terrain model of 5th generation (size of the area is 680 × 520 m).

4 Project of Temporary Bridge TMS

The main task for reconnaissance team (reco-team) was the reconnaissance of area where the bridge was planned to be built. This team consists of experts from Military Geodetic and Hydrometeorological Office from Dobruška and one member from Department of Engineer Technology who has experience with TMS set. Geodesists were responsible for geodetic measurement of the construction site of temporary bridge. The reco-team carried out inspection of the construction site and made photo documentation. They found out span of the obstacle and suggested which type of bridge is suitable for particular gap. All information was included in Reconnaissance Report. When the reco-team accomplished the report, they placed it on IPEC. The geodesists were led by commander of the reco-team and under his control they made detailed geodetical survey. The data gained from survey were essential part of the Reconnaissance Report and were attached to the report as soon as geodesists prepared them. The data consisted of coordinates of important points and construction site drawing in MicroStation software [12].

As a suitable temporary bridge for overbridging river was chosen TMS bridge set. The TMS Bridge set is the main representative of temporary bridges used for logistic. TMS is road, standardized steel bridge. This bridge is designed for only one traffic lane (4 m wide). The TMS Bridge is a through-type truss bridge, the roadway being carried between two main girders. The trusses in each girder are formed by 3 - m panels pinned end to end. It is easy to install. Each part of the TMS Bridge is a standard machine-made piece and is interchangeable among spans. In few cases (single story), no heavy equipment is required to assemble or launch a TMS Bridge; only basic pioneer skills and equipment are needed. TMS Bridge is determined for wide span (up to 36 m for MLC 70). It is the most suitable to build one bridge span but it is also possible to build bridge with more spans (max 7 spans). In this case is necessary to build pier (e.g. PIŽMO). For the construction of pier is also possible to use alternative material (pile bent, grillage using timber or concrete panels) [13] (Fig. 3).

Fig. 3. Model of bay TMS and MS bridge set created by Autodesk Inventor SW [12].

The main girders on each side of the centerline of the bridge can be assembled from a single truss or from two or three trusses side by side. For greater strength, a second story of panels can be added to the trusses. The upper stories are bolted to the top chord of the lower story. Transverse floor beams, called transoms, are clamped to the bottom chords of the trusses and support stringers and decking. Sway braces between the girders provide horizontal bracing and transoms keep the trusses upright. The bracing frames between the trusses provide lateral bracing within each girder. Load carrying capacity depends on span of the bridge and type of the construction [13]. This bridge set is not equipment of Czech Army. The bridge set is in storage of Ministry of Transport. The soldiers of Engineer Corps are trained in special courses in Training center of Ministry of Transport, where they get the experience with this bridge construction and also with other bridge constructions.

The original bridge was damaged by demolition beams and deck. Temporary bridge construction (TMS) was designed outside the axis of the original bridge. On the right river bank could exploit an existing road. Any driveway did not exist on the other side of the river. There had to design new temporary road. The width of the flow was of 163 m and width of the channel 172 m. Most of structure TMS is provisional, demountable, joist, truss, with the lower deck. The bridge is open configuration. The main beams are assembled from prefabricated and length of 3 m. The supporting structure of the bridge is typed; the basic element is mounting compartment. Total length of temporary bridge was designed 180 m (39 + 33 + 36 + 33 + 39), road width between curbs 4.0 m. Bridge had no sidewalks. Foot traffic on the bridge is possible to implement measures to prevent the collapse of the bridge (e.g. mesh, wooden guard-rails etc.) Carrying capacity of the bridge was under the [14] determined as follows:

- Normal capacity: 13 tons
- Load per axle: 11 tons
- Caution capacity: 38 tons
- Risk capacity: to prove it static calculation

The leader of project team specified type of construction, site for construction and span of the bridge. Team created 3D model of bridge construction (Fig. 4). This 3D model was also used for making project documentation. Designing necessary ground shaping in area of construction was other part of project. Microstation situation data were converted into proper CAD format. Then team created into terrain ground shaping and other modification. Then the 3D model of bridge was placed into model of terrain. The last part of designing was creating the final project and drawing documentation. The final project with drawing was placed on the Information Portal of Engineer Corps (IPEC, see more in [15, 16]).

Fig. 4. Model of the bridge in photo.

5 Project of Temporary Road

The temporary bridge project related to its connection to the existing road network. Firstly was an engineer reconnaissance of both river banks carried. As stated above, on the right river bank was possible to use existing roads (roads lead up to the place of construction of temporary bridge). The situation on the left river bank was a little complicated. It was necessary to design two new temporary roads, which connected temporary bridge with local road in the village České Kopisty. Total length of these temporary roads was 1,462 m (one-lane roads). New elevation data of the Czech Republic was the basis for temporary roads designing. The main base was the DMR 5G. DMR 5G for the locality was exported in four text files (ASCII encoding). Each file contained a locality 2 × 2 km, which means about 400,000 points in a single file, each in size from 10 to 14 MB (based on points). An Orth photo of the area was also used for the designing.

DTM was created from the text files (encoding ASCII), which contained x-, y-coordinates and altitudes of points. Initially, these points were imported into AutoCAD Civil 3D program and there was created DTM, which was displayed by TIN. Next the direction, the vertical alignment and the sample cross-sections of the road were designed. AutoCAD Civil 3D generated automatically the road corridor and eventually the table of earthworks volumes was made (Fig. 5).

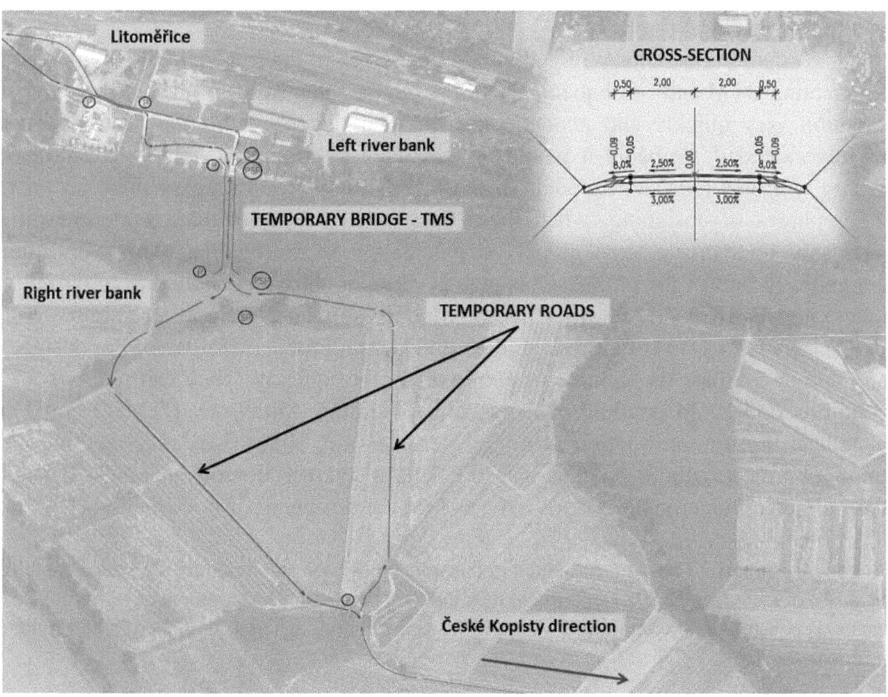

Fig. 5. Situation of the temporary roads.

The whole design of temporary roads was created by Autodesk's software AutoCAD Civil 3D. AutoCAD Civil 3D software is Autodesk's building information modelling solution for civil engineering. AutoCAD Civil 3D produces a single model, with intelligent and dynamic data, enabling designers to more quickly make a design change at any stage of the process. The model automatically reflects any changes to drafting and annotation throughout the project. The software enables designers to more quickly process earth volumes between existing and proposed surfaces [17]. This sophisticated software offers various functions. Only some of them have been used, especially Generation DTMs from survey data and corridor modelling, which combines horizontal and vertical geometry with flexible cross sectional components to create a parametrically defined dynamic 3D model of the roads.

6 Conclusion

Joint exercises Surveyors and Engineers confirmed the achieved abilities of the individual components. On the basis of geodetic (focus obstacles In Situ elevation of new models, DTM5G) was elaborated bridge replacement project, including temporary access roads. Processing project of bridge's provisory took approximately 16 h to process both temporary roads for about 14 h. These times are approximate (depending on local site conditions and the skill of the designer), can be the basis for the planning of forces and resources in crisis situations. Topic broken bridge under the critical transport infrastructure is a typical case of an emergency and you need to be to tackle well prepared.

A geodetic measurement directly done in the area of interest is still the main source for creating DTM and subsequent designing. On the other hand, we have to solve crisis situation very quickly and geodetic measurement in situ could the design process rapidly extended. Probably it would be beneficial to combine geodetic measurement in situ with new elevation models of the Czech Republic. Moreover there are many other phases of designing, when these data could be exploitable. Sometimes it is necessary to estimate very quickly a calculation of earthworks, which constitute a significant part of the project. This task is very common especially during designing temporary roads or bridges after floods. Modern software offers a very simple comparison between DTMs before and after flood. Using modern software, the designers are able to estimate the volume of earthworks more quickly. They can assess bank's condition and avoid possible dangers before building. Moreover, DTM5G could be beneficial during construction planning, designers are able to get very credible preliminary project without geodetic measurement in situ. It is obvious that these data can help to solve numerous problems even before the designing of constructions.

Acknowledgment. The work presented in this article has been supported by the Czech Republic Ministry of Defence - K 201 Department development program "Development of technologies in weapons construction, ammunition, weapons instrumentation, materials engineering and military infrastructure".

References

1. Czech Republic. The law on the defense of the Republic. J. Laws, **2010**(222) (1999)
2. Czech Republic. The law on crisis management and amending certain Acts (Crisis law), as amended. J. Laws, **2010**(240), No 149 (2000)
3. Benda, M., Sobotka, J.: Utilize autocad civil 3D during designing of temporary bridges projects. In: ICMT'11 - International Conference on Military Technologies 2011, University of Defence, Brno, pp. 231–236 (2011). ISBN 978-80-7231-787-5
4. Maas, H.-G.: Akquisition von 3D-GIS Daten durch Flugzeuglaser-scanning. Kartographische Nachr. **55**(1), 3–11 (2005)
5. Kraus, K., Pfeifer, N.: Determination of terrain models in wooded areas with airborne laser scanner data. ISPRS J. Photogrammetry Remote Sens. **53**(4), 193–203 (1998). doi:10.1016/S0924-2716(98)00009-4. ISSN 0924-2716, http://www.sciencedirect.com/science/article/pii/S0924271698000094
6. Bartels, M., Wei, H.: Threshold-free object and ground point separation in LIDAR data. Pattern Recogn. Lett. Pattern Recogn. Remote Sens. **31**(10), 1089–1099 (2010). doi:10.1016/j.patrec.2010.03.007. ISSN 0167-8655, http://www.sciencedirect.com/science/article/pii/S0167865510000887
7. The project of creating a new elevation of the Czech Republic (in Czech). Land Survey Authority, Praha (2008)
8. Brázdil, K.: The project of creating a new elevation of the Czech Republic (in Czech). In: Proceedings - Symposium GIS Ostrava 2010: GIS meets Remote Sensing and Photogrammetry towards Digital World. Ostrava: VSB - Technical University of Ostrava campus, pp. 72–79 (2010). ISBN 978-80-248-2171-9, ISSN 1213-239X. http://gis.vsb.cz/GIS_Ostrava/GIS_Ova_2010/sbornik/Lists/Papers/CZ_5_2.pdf
9. Hubáček, M., et al.: Accuracy of the new generation elevation models. In: ICMT 2015 - International Conference on Military Technologies, Brno, Czech Republic, pp. 1-9 (2015)
10. Hubáček, M., Čeplová, L., Břeňová, M., Zerzán, P.: Verification of accuracy of the new generation elevation models. Adv. Mil. Technol. **9**(2), 21–32 (2014). http://aimt.unob.cz/articles/14_02/14_02%20(2).pdf, ISSN 1802-2308
11. Hubáček, M., Kovařík, V., Kratochvíl, V.: Analysis of influence of terrain relief roughness on DEM accuracy generated from lidar in the Czech Republic territory. Int. Arch. Photogrammetry, Remote Sens. Spat. Inf. Sci. **XLI-B4**, 25–30 (2016). doi:10.5194/isprs-archives-XLI-B4-25-2016
12. Benda, M.: Construction of tms bridge in Víska village – Example of temporary replacement. In: Security Management And Society, Univerzita obrany, Brno, pp. 32–39 (2012). ISBN 978-80-7231-871-1
13. Maňas, P., Soušek, R., Benda, M., et al.: Temporary bridge construction from TMS set, Institute Jana Pernera, Brno, o.p.s., 120 p (2010). ISBN 978-80-86530-73-4
14. Technical conditions of use temporary bridges from TMSset in the civil sector. MDS – OPK 24911/96-120
15. Maňas, P., Mazal, J.: The reach-back concept in the Czech army corps of engineers. In: International Conference on Military Technologies, Faculty of Military Technology, University of Defence, Brno (2009). ISBN 978-80-7231-648-9

16. Maňas, P., Soušek, R.: On Cooperation between military and civilian authorities in the Czech republic during crisis situation in transport. In: International Conference on Engineering and Meta-Engineering icEME 2010, Orlando, Florida, USA, IIIS, International Institute of Informatics and Systemics, pp. 12–14 (2010). ISBN 978-1-934272-83-1
17. Civil-3D. Autodesk. Autodesk, © 2016. Accessed 19 Oct 2016. http://usa.autodesk.com/civil-3d/features/all

Engineer Construction of Provisional Bridges in the Army of Czech Republic

Jindřich Holopírek(✉) and David Lysoněk

FVT, University of Defence Brno, Kounicova 65, 66210 Brno, Czech Republic
{jindrich.holopirek, david.lysonek}@unob.cz

Abstract. This article analyzed the issue of provisional bridges engineer superstructures - mechanized bridges in the Army of the Czech Republic (ACR). Current devices are outdated and their parameters no longer correspond to the real needs of our military. Mechanized bridges are devices for rapid and effective security units move through impassable places (water obstacles, trenches, walls, etc.). They are designed not only for ACR units but also for transport means for intervention in emergency situations (floods, and other natural disasters) within the Integrated Rescue System (IRS). The article shows one of the options approach to modernization solutions to this type of provisional bridges, which is based on an analysis of current and real needs of the ACR.

Keywords: Mechanized bridge · Bridge · Obstacle

1 Introduction

Mechanized bridges are designed for engineer support movement combined arms units. These are bridges which are transported on a wheeled or tracked chassis and these carriers are considered to be an obstacle. These bridges provide quick bridging barriers within minutes.

These mechanized bridges were introduced in the 80's to the Army of the Czech Republic. Parameters, namely tonnage and length of the bridge reflect the needs of the former army. The structure and determine current ACR has changed dramatically, but the existing mechanized bridges were not upgraded. Modernization provisional bridges should be based on current requirements ACR and a thorough analysis not only territory where it is expected to deploy but also on the analysis of means of transport for which they are intended.

2 Current Mechanized Bridges in ACR

2.1 AM-50

The armament of the ACR bridge automobile AM-50 introduced in 1975. The actual deck is on the chassis of the T 813 8 × 8. The bridge is a scissor structure, has a length of 13.5 m and is designed to load 50 t. It is a bridge on solid supports which can be disconnected as needed or used for support in a barrier. Preferably the joining to 8 parts (Fig. 1).

A. Kravcov et al. (eds.), *Durability of Critical Infrastructure, Monitoring and Testing*,
Lecture Notes in Mechanical Engineering, DOI 10.1007/978-981-10-3247-9_21

Fig. 1. AM-50

2.2 MT-55, PM-55

Bridge tank MT-55, is intended to secure the movement of particular tank units. The bridge was introduced into the ACR in 1973. The deck is located on a tank chassis T-55, which is placed with the bridge laying equipment scissor structure with a total length of 18 m. To secure a replacement bridge for the MT-55 is used transporter bridge PM-55 which is located on the T-813 8 × 8 (Figs. 2 and 3).

Fig. 2. MT-55

Fig. 3. PM-55

2.3 AM 70

The latest mechanized type bridge the AM-70, designed to load MLC 70. This mechanized bridge based on the principle of the bridge structure AM 50. Unfortunately it was a botched project, and not introduced in the ACR (Fig. 4).

Fig. 4. AM-70

3 Problem Definition

The current structure of the ACR consists mainly of light motorized units, medium and mechanized units for rapid deployment. The vast majority of logistic support is provided by means of transport to a maximum weight of 30 tons. Deployment of Army units within the IRS, for example by ensuring provisional bridges for intervention in vulnerable areas, does not require a load of these bridges over 30 tons. Quantity mechanized bridges that can currently be used is very small. Hence the need to address this issue by modernizing provisional bridges - mechanized bridges for the current real needs of the ACR.

4 The Suggestion of Solution

Draft mechanized bridge to ensure the real needs of today's military influence mainly these factors. Firstly, please specify which technology will use mechanized bridges, both for security support engineer movement combined arms units and logistic support and when possible deployment within the IRS. Another factor is to analyze the space in which it is expected to deploy. A necessary step for the choice of the appropriate knowledge of the existing solutions - a detailed research and its analysis. Design your agent can then deal with your own design or buy an existing one.

1. **Analysis techniques for which it is intended mechanized bridge**

One of the main factors influencing the choice of the percentage of techniques in the ACR. Graph 1 shows the percentage of the number of technology in our army. The graph shows that the largest percentage of technology in our army occupied Tatra T-815, then UAZ, Land Rover and finally Tatra T-810 and T-148th.

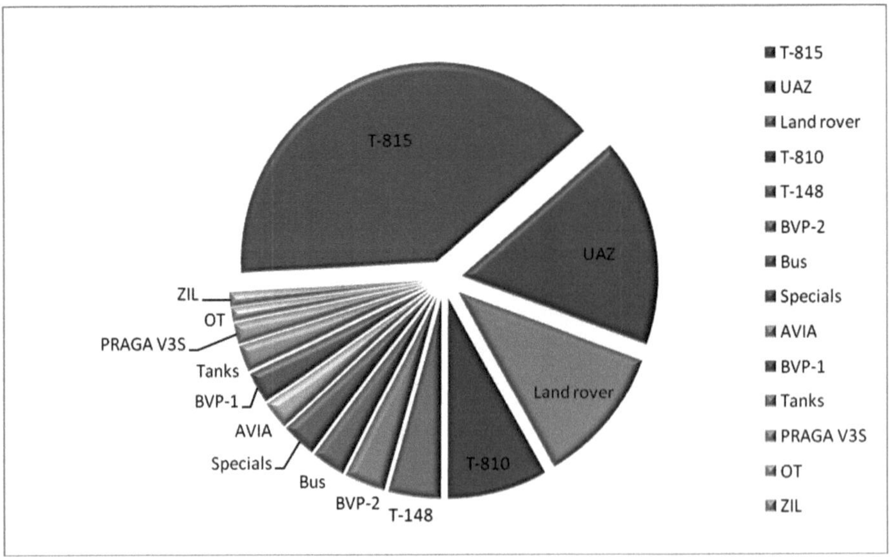

Graph 1. Percentage distribution techniques in the ACR

If the technology divide by weight, may be in Graph 2 to determine: In the category of up to five tonnes, 28.21% of the vehicles in the category from five to ten tons - 3.39%, from ten to fifteen tons - 26.80% since fifteen to twenty tons - 39.49% of the total amount of vehicles located in the ACR. The lowest percentage is taken by the heaviest category of vehicles and the means weighing from twenty to fifty tons, and these funds is only 2.12% of the total number of vehicles in the military.

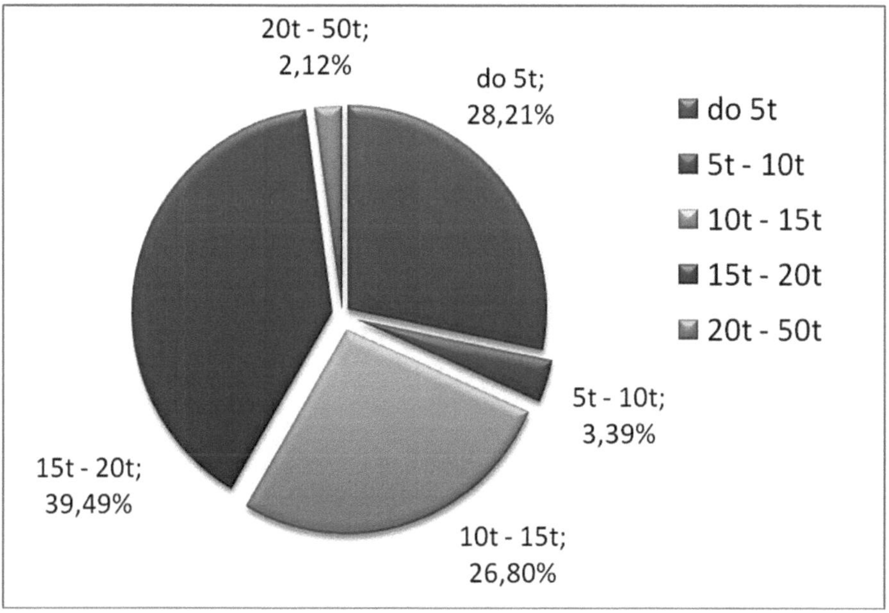

Graph 2. Allocation of vehicles into five classes according to weight

If the technique is divided into three weight groups can Graph 3 to determine its frequency:

- 2.12% - funds from twenty to fifty tons,
- 31.59% - vehicles weighing up to ten tons,
- 66.29% - is in the range of from ten to twenty tons.

An important factor in selecting mechanized bridges is also possible use in deployment within the IRS in emergencies (floods, etc.). It can assume that the establishment of provisional bridges using mechanized bridges will be crucial to their swift establishment. Technique for initial intervention in vulnerable areas in the vast majority has no mass greater than 30 tons.

2. **Analysis of the area of deployment of mechanized bridges**

Mechanized bridges are designed mainly for the establishment of a bridge over a water hazard. If we analyze e.g. the geographical distribution of rivers in the Czech Republic, you can Graph 4 to determine:

- 50% of water hazards in the Czech Republic are the width 3 m,
- 25% are obstacles to a width of 6 m
- 15% are obstacles to a width of 15 m
- less than 10% are obstacles to a width of 50 m.

It follows that in the area of the Czech Republic is often needed to overcome the water barrier to the length of 10 m.

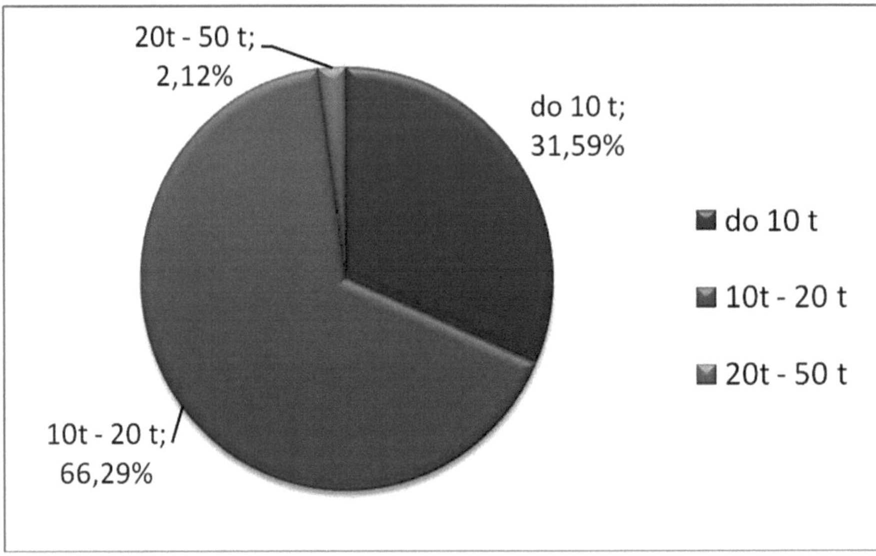

Graph 3. Allocation of vehicles into three classes according to weight

Deployment mechanized bridges, it is possible not only when needed to overcome the water barrier, but also non-aqueous barrier. The following graph shows the cumulative frequency of most obstacles. The Graph 5 can then specify that 60% of barriers in Europe is in the width of 12 m.

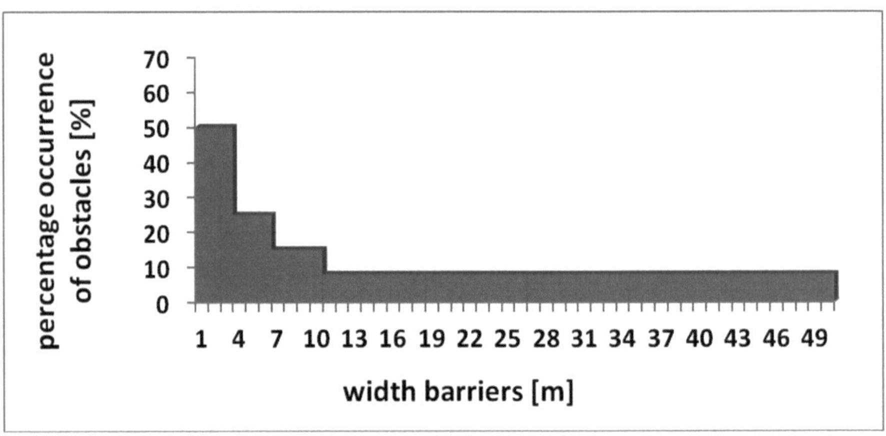

Graph 4. The frequency of water hazards in the Czech Republic in accordance with their width

3. Search of existing mechanized bridges

Security forces move mechanized bridges solves all the armies and have their own resources. Knowledge of this technology, its principles and parameters necessary for

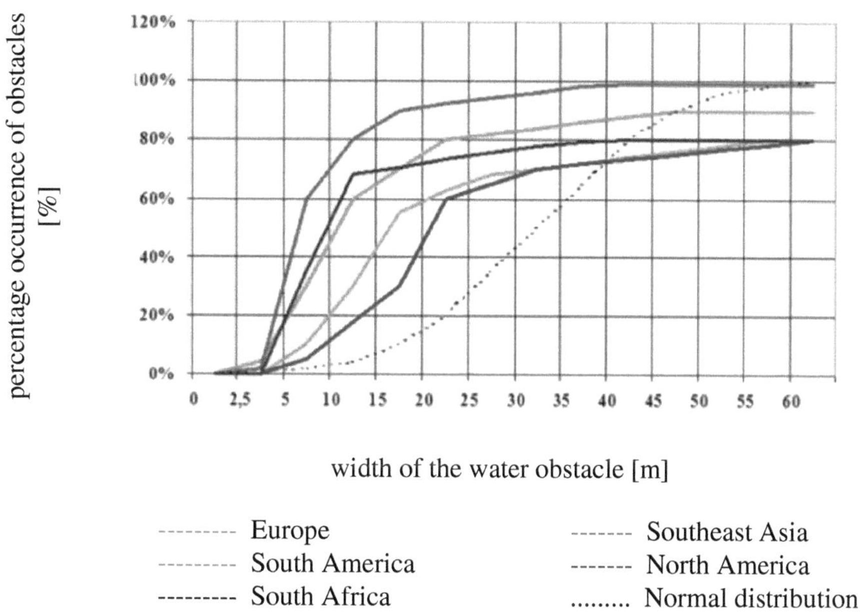

Graph 5. Cumulative frequency barriers

the proposal to modernize mechanized bridges for ACR. Using some of the methods of multi-criteria decision-making to determine what type of existing mechanized bridges would be suitable for the conditions of the ACR.

4. **Proposal mechanized bridge ACR**

Proper design of mechanized bridge for the needs of the ACR can be solved by buying appropriate, existing mechanized bridge from another state or by the structural design, which respects fundamental factors mentioned above.

5 Conclusion

From the above analysis it is clear that mechanized bridges in the ACR for the present unsatisfactory conditions. The complete absence of mechanized bridges in light mechanized units, which require lightweight bridges with a length of at least 12 m to quickly overcome obstacles. It is proposed bridges on wheeled vehicles and located in such units. A suitable solution would be to place the layup device with a bridge on the technology introduced with these units.

An important factor is also our integration within NATO. Each army should be able to secure the movement of their own funds. It can be expected in the event of deployment of our troops in the international groupings that each army has its technique for adequate means to overcome obstacles. Modernization mechanized bridges ACR must therefore respect the particular requirements of their own units.

The issue of water and non-aqueous overcoming obstacles for the infantry in the Army is also not sufficiently addressed. There is only a bridge for EOD units. The need for lightweight ladder to overcome obstacles to a minimum width of 12 m confirmed by the experiences of foreign armies, and also has experience in dealing with emergencies during the deployment of the army within the IRS.

It can conclude that the modernization of mechanized bridges for the needs of the ACR is necessary to focus primarily on issues of overcoming obstacles for medium and light mechanized units.

Acknowledgment. The work presented in this article has been supported by the Czech Republic Ministry of Defence - K 201 Department development program "Development of technologies in weapons construction, ammunition, weapons instrumentation, materials engineering and military infrastructure".

References

1. Lysoněk, D.: Návrh vhodných typů mechanizovaných mostů pro podmínky AČR. Brno, 2016. Bakalářská práce. Univerzita obrany. Vedoucí práce Ing. Zdeněk Hejmal, CSc (2016)
2. Jane's Military Vehicles and Logistics 2011-2012. 2012. Janes Information Group (2011). ISBN 978-071-0629-524
3. Kohoutek, J.: Katalog ženijní techniky a materiálu EOD. Praha: Ministerstvo obrany České republiky - AVIS, 163 s (2006). ISBN 80-727-8315-7
4. TTE-151451/SER:NU0083. BI-SC Capability Codes and Capability Statements. North Atlantic Treaty Organization (2016)
5. Český obranný standard: Mechanizované mosty všeobecné technické požadavky. In: Praha: Vojenský technický ústav ochrany Brno, 2005, ročník 2005, číslo 999917. Dostupné také z. http://www.oos-data.army.cz/cos/cos/999917.pdf

The Dynamic Modulus of Elasticity as an Important Parameter for Military Use of Constructions

Eva Zezulová(✉)

Faculty of Military Technology, University of Defence,
Kounicova 65, 66210 Brno, Czech Republic
eva.zezulova@unob.cz

Abstract. The article deals with the importance of the dynamic modulus of elasticity for military use of protective constructions. Not only the construction of the shelter has to save the crew inside, but also it should bear loading repeatedly. And the verification procedures of construction capability to withstand further load are being developed by military engineer specialist for the Czech armed force's needs.

But for the verification of the real shelters construction is not possible to take samples. That's why the non-destructive methods are used. From these methods is possible to obtain the dynamic modulus of elasticity and the static modulus has to be counted.

The paper contains the results of measurements of the dynamic modulus of elasticity with ultrasound to test of the 20 cubes made from the concrete and 18 cubes made from the steel fiber reinforced concrete. The dynamic modules of elasticity were converted to the static in accordance with Czech technical standard ČSN 73 2011 [2]. Non-destructive testing of concrete structures.

Keywords: Dynamic modulus of elasticity · Static modulus of elasticity · Military constructions

1 Introduction

Protective constructions are an integral part of military equipment from ancient history to the present. Protective constructions are mainly used to protect the live force. Their main task is to minimize the loss of soldiers filling task lives. During the times required characteristics of protective structures have changed depending on the weapons type against which they are protecting [1, 3]. Currently, protective structures must withstand the effects of the pressure wave, explosion and penetration. For these types of loads is very important parameter which is the modulus of elasticity of the material [1, 3, 4].

Reinforced concrete is the traditional material, which provides sufficient shelter for troops. Simultaneously with the development of weapons ere developed new materials which are capable of withstanding these new weapons [5, 6]. The steel fiber reinforced concrete seems to be a very suitable material because of its suitable characteristics [7]. It is a material with high strength can maintain a high modulus of elasticity.

A. Kravcov et al. (eds.), *Durability of Critical Infrastructure, Monitoring and Testing*, Lecture Notes in Mechanical Engineering, DOI 10.1007/978-981-10-3247-9_22

The modulus of elasticity is one of the basic characteristics of concrete [3, 8, 9]. Significantly it affects the deformation properties of concrete and hence deformation of concrete structures such as deflections, displacements, loss and so on. In general, the larger the elastic modulus of concrete is, the smaller the deformations are [10, 11]. And vice versa. With the arising of modulus of elasticity, the concrete strains and reduces the deflection of reinforced concrete structures are falling [12].

The modulus of elasticity importance as one of the basic characteristics of concrete increases with the static demands of structure, with a growth of construction loading. Such a demanding load is an explosion or projectile penetration of the shelters construction [12].

Not only the construction of the shelter has to save the crew inside, but also it should bear loading repeatedly. And the verification procedures of construction capability to withstand further load are being developed by military engineer specialist for the Czech armed force's needs.

2 The Static Modulus of Elasticity

The modulus of elasticity can be defined as the ratio of tension and relative deformation caused by him, which can be expressed by the equation:

$$E = \frac{\sigma}{\varepsilon} \tag{1}$$

Where:

E The modulus of elasticity [MPa],
σ Stress [MPa],
ε Relative deformation

Static modulus of elasticity in compression, according to ČSN ISO 1920-10 [13] determines the deformations that occur when a known load [15, 16]:

$$E_C = \frac{\Delta\sigma}{\Delta\varepsilon} = \frac{\sigma a - \sigma b}{\Delta\varepsilon} \tag{2}$$

Where:

σa top loading stress [N/mm^2];
σb basic loading stress [N/mm^2];
Δε the average change of relative deformation between the top and the base tension.

3 The Dynamic Modulus of Elasticity

In addition to the static modulus of elasticity, which is determined from the real stress and deformation of the sample subjected to stresses, there is another type of The modulus of elasticity, which is called the dynamic modulus of elasticity.

Dynamic elastic modulus can be determined by several methods, the most common include resonance and ultrasound, method.

The resonance method is based on the measurement of the first natural frequency of vibration of the test sample. Dynamic elastic modulus of the resonance frequencies in [N/mm^2] can be set in two ways; from the longitudinal vibration as dynamic elastic modulus $E_{dyn,FL,}$ and from the transverse vibration as dynamic elastic modulus $E_{dyn,FF}$. [9]

From the longitudinal vibration:

$$E_{dyn,FL} = 4 \cdot L^2 \cdot f_L^2 \cdot D \tag{3}$$

From the transverse vibration:

$$E_{dyn,FF} = 0.0789.c.L^4.f_f^2.b.i^{-2} \tag{4}$$

Where:

f_L, f_f natural frequency of longitudinal and transverse vibrations [kHz];
L sample length [m];
D density of concrete [kg/m^3];
i radius of inertia;
c value depends on ratio of $\frac{i}{L}$ and on Poisson's ratio. μbr

For ultrasound method is value of the dynamic modulus of elasticity dynamic modulus of elasticity in compression and tension counted from equation [6]:

$$E_{bu} = \rho \cdot \frac{v_L^{2.1}}{k_3^2} \cdot 10^{-6} \tag{5}$$

Where:

ρ Density [kg.m^{-3}];
v_L Pulse velocity of longitudinal ultrasonic waves [m.s^{-1}];
k_3 The factor of environment dimensionality (k_1,k_3) see Table 1

The standard ČSN 73 1371 discloses a method for determining of coefficients k_2 and k_3 for two-dimensional and three-dimensional environment. These values are dependent on the Dynamic Poisson Coefficient v_{bu} and they are determined using equations [17]:

$$k_2 = \sqrt{\frac{1}{1 - v_{bu}^2}} \tag{6}$$

$$k_3 = \sqrt{\frac{1 - v_{bu}}{(1 + v_{bu}).(1 - 2v_{bu})}} \tag{7}$$

Table 1. Coefficients k_2 and k_3 [17]

v_{bu}	*k2*	*k3*	*k3/k2*
0,00	1,0000	1,0000	1,0000
0,02	1,0002	1,0004	1,0002
0,04	1,0008	1,0017	1,0009
0,06	1,0018	1,0039	1,0020
0,08	1,0032	1,0070	1,0038
0,10	1,0050	1,0113	1,0062
0,12	1,0073	1,0168	1,0094
0,14	1,0100	1,0236	1,0135
0,16	1,0130	1,0319	1,0187
0,18	1,0166	1,0420	1,0250
0,20	1,0206	1,0541	1,0328
0,22	1,0251	10,685	1,0423
0,24	1,0301	1,0857	1,0539
0,26	1,0356	1,1061	1,0681
0,28	1,0417	1,1307	1,0854
0,30	1,0483	1,1602	1,1068
0,32	1,0555	1,1963	1,1334
0,34	1,0633	1,2406	1,1667
0,36	1,0719	1,2964	1,2095
0,38	1,0811	1,3682	1,2656
0,40	1,0911	1,4639	1,3416
0,42	1,1019	1,5978	1,4500
0,44	1,1136	1,8002	1,6168
0,46	1,1262	2,1502	1,9092
0,48	1,1399	2,9637	2,6000
0,50	1,1547	∞	∞

For one-dimensional environment is the coefficient:

$$k_1 = 1 \quad (8)$$

4 The Evaluation of the Dynamic Modulus of Elasticity on the Testing Samples

All concrete and steel fibre concrete cube-shaped samples were non-destructively tested using ultrasound before their destructive testing. The measured values were used to calculate the dynamic modulus of elasticity. For your own calculation equation was used Eq. 5 according to ČSN 73 1371 [17]. The coefficient k_3 dimensionality of the environment was determined from Table 2 from ČSN 73 1371 [17]. For the dynamic Poisson coefficient of concrete v_{bu} = 0.20 was determined value 1.0541 as the coefficient table dimensionality of the environment k_3.

Table 2. Measured values of ultrasound waves in points 1–3 (Fig. 1) and calculate dynamic modulus of elasticity of concrete according to Eq. 5

I. D.	Sample dimensions [mm]			Weight	Density	Speed of ultrasound waves [m.s^{-1}]			Diameter for point 1–3	Dynamic modulus of elasticity
	a	b	c	[kg]	kg/m^3	point 1	point 2	point 3	[m.s^{-1}]	[MPa]
1	148.0	147.5	150.0	8.572	2618	4850	4800	4800	4817	54659
2	149.0	149.5	149.8	8.586	2573	4860	4790	4790	4813	53653
3	149.8	148.2	149.6	8.564	2579	4840	4690	4690	4740	52142
4	149.6	149.6	149.7	8.585	2563	4730	4870	4930	4843	54100
5	149.7	148.8	149.7	8.570	2570	4820	4810	4700	4777	52777
6	150.2	149.8	149.6	8.689	2581	4820	4870	4720	4803	53599
7	147.8	149.5	149.6	8.636	2613	4810	4820	4790	4807	54326
8	149.8	150.0	149.2	8.603	2566	4830	4730	4690	4750	52109
9	149.7	149.6	149.4	8.500	2540	4570	4720	4790	4693	50362
10	151.3	149.5	149.6	8.666	2561	4690	4660	4560	4637	49551
11	150.6	149.8	149.9	8.576	2536	4620	4700	4730	4683	50059
12	149.7	150.0	151.0	8.647	2550	4700	4780	4720	4733	51419
13	150.2	150.1	150.2	8.652	2555	4590	4720	4780	4697	50724
14	149.8	150.2	149.7	8.621	2559	4750	4700	4720	4723	51389
15	150.1	149.8	149.9	8.568	2542	4550	4570	4570	4563	47642
16	149.4	151.2	149.5	8.654	2562	4750	4720	4750	4740	51815
17	152.2	149.6	149.8	8.742	2563	4620	4730	4790	4713	51244
18	149.5	149.6	149.7	8.618	2574	4760	4750	4690	4733	51902
19	149.5	149.6	149.7	8.728	2607	4600	4560	4750	4637	50436
20	149.6	149.7	150.2	8.624	2564	4700	4620	4490	4603	48895

The dynamic modulus of elasticity of concrete we can find by using statistics as the average value of the dynamic modulus of elasticity. The dynamic modulus of elasticity in this case is E_{bu} = 516,4 GPa. The dynamic modulus of elasticity can be converted to static modulus of elasticity using the formula (9) according to ČSN 73 2011 [2].

$$E_{bs} = \frac{\sum E_{bu}}{n} = \frac{E_{bu1}}{1}\chi_u = E_{bu,měě} \cdot \chi_u \tag{9}$$

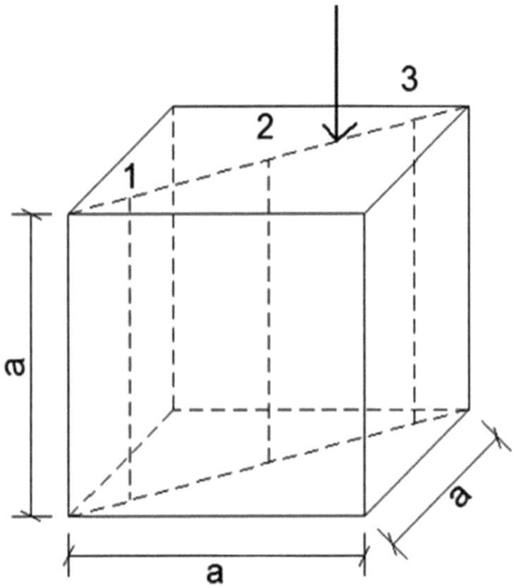

Fig. 1. Points 1-3 the places where the measured values of ultrasound waves were taken

where:

E_{bu} the dynamic modulus of elasticity of concrete determined according to ČSN 73 13 71 [10] [GPa],

n the number of test sites,

χ_u reducing coefficient in accordance with Table 3

Table 3. The reducing coefficient [17]

The strength class of concrete	The reducing coefficient X_u (for the ultrasound method)	The reducing coefficient X_r (for the resonance method)
C8/10	0,62	0,81
C12/15	0,71	0,86
C16/20	0,76	0,88
C25/30	0,81	0,90
C30/37	0,83	0,91
C35/45	0,86	0,93
C40/50	0,88	0,94
C45/55	0,90	0,95

The dynamic modulus converted to the static modulus of elasticity of concrete:

$$E_{bs} = E_{bu,měě} \cdot \chi = 51,6 \cdot 0,96 = 49,6\,GPa \tag{10}$$

The dynamic modulus of elasticity calculated on the static modulus of elasticity of concrete:

$$E_{bs,\text{steel fibre concrete}} = E_{bu,m\check{e}\check{e}} \cdot \chi_u = 48,9 \cdot 0,97 = 47,4\,GPa \tag{11}$$

Values of the reducing coefficient χ_u = 0,96 for concrete and χ_u = 0,97 for steel fibre concrete were values of coefficient extrapolated from the Table 3 [6]. Exploitation values for the fiber reinforced concrete was necessary because the values in Table 4 are only to the strength of 60 MPa.

Table 4. Measured values of ultrasound waves at spots 1–3 and calculation of the fiber reinforced concrete dynamic modulus of elasticity according to the Eq. 5

I. D.	Sample dimensions [mm]			Weight	Density	The speed of ultrasound waves [m.s^{-1}]			Diameter of values 1-3	Dynamic modulus of elasticity
	a	b	c	[kg]	kg/m^3	bod 1	bod 2	bod 3	[m.s^{-1}]	[MPa]
22	149.3	150.2	149.8	8.720	2596	4530	4700	4690	4640	50295
23	150.0	149.8	151.3	8.655	2546	4570	4570	4560	4567	47779
24	149.6	150.3	150.0	8.597	2549	4620	4600	4530	4583	48193
25	149.5	150.3	159.6	8.573	2391	4550	4660	4520	4577	45064
26	150.0	150.0	151.8	8.621	2524	4570	4570	4490	4543	46891
27	149.2	149.5	151.6	8.610	2546	4480	4640	4620	4580	48067
28	149.4	149.7	150.1	8.612	2565	4620	4650	4650	4640	49708
29	149.9	149.8	150.2	8.654	2566	4570	4620	4620	4603	48936
30	150.1	150.2	149.7	8.697	2577	4620	4610	4570	4600	49074
31	149.6	149.7	149.9	8.598	2561	4580	4600	4550	4577	48282
32	149.9	150.0	150.3	8.673	2566	4600	4620	4620	4613	49159
33	150.2	150.1	150.1	8.713	2575	4690	4700	4640	4677	50683
34	149.7	149.8	150.1	8.690	2582	4650	4620	4650	4640	50025
35	149.8	149.7	150.1	8.720	2591	4670	4650	4650	4657	50558
36	149.8	149.6	150.1	8.654	2573	4580	4620	4600	4600	48995
37	149.7	149.9	150.2	8.679	2575	4580	4750	4730	4687	50903
38	149.6	149.8	150.1	8.690	2583	4620	4680	4680	4660	50491
39	150.1	150.0	150.2	8.720	2579	4580	4530	4550	4553	48114

5 Summary

Not only the construction of the shelter has to save the crew inside, but also it should bear loading repeatedly. And the verification procedures of construction capability to withstand further load are being developed by military engineer specialist for the Czech armed force's needs.

But for the verification of the real shelters construction is not possible to take samples. That's why the non-destructive methods are used [8]. From these methods is possible to obtain the dynamic modulus of elasticity and the static modulus has to be counted.

Whereas, in the non-destructive determination The modulus of elasticity it is not exposed to the real load of concrete and thus in the sample there are no tension there or in the creation of micro cracks at the interface of aggregate and cement stone.

The advantage of dynamic modulus of elasticity is the speed and ease of setting. The disadvantage of a lower accuracy and the reliability of measured values in comparison with the static modulus of elasticity.

The paper contains the results of measurements of the dynamic modulus of elasticity with ultrasound to test of the 20 cubes made from the concrete and 18 cubes made from the steel fiber reinforced concrete. The dynamic modules of elasticity were converted to the static in accordance with Czech technical standard ČSN 73 2011. Non-destructive testing of concrete structures [7].

Based on the results of the experimental measurements of concrete and fibre concrete samples were calculated static modules of elasticity for concrete and for fibre reinforced concrete. For samples made from the plain concrete is E_{bs} = 49.6 kPa, and static modulus of elasticity of the fibre reinforced concrete for the $E_{bs,\ steel\ fiber\ reinforced\ concrete}$ = 47.7 GPa.

Acknowledgment. The work presented in this article has been supported by the Czech Republic Ministry of Defence- K 201 Department development program "Development of technologies in weapons construction, ammunition, weapons instrumentation, materials engineering and military infrastructure".

References

1. Štoller, J., Dvořák, P.: Field tests of high performance fiber reinforced concrete slabs: Impact of contact and distant explosions. In: 2015 International Conference on Military Technologies ICMT 2015, art. no. 7153679 (2015). ISBN: 978-807231976-3, doi:10.1109/MILTECHS.2015.7153679
2. Czech technical standard ČSN 73 2011. Non-destructive testing of concrete structures, 23 pages. ÚNM Publishing, Prague (2012)
3. Maňas, P., Balogh, Z.: Critical Infrastructure Protection - Principles and Practical Approaches, pp. 30–35. Counter-IED Report, 2016, Spring/Summer (2016). ISSN 2050-6732
4. Štoller, J., Maňas, P., Zezulová, E.: Blast testing and simulation methods, pp. 35–100. Czech Technology -Publishing House ČVUT - Prague, Praha (2015). ISBN 978-80-01-05898-5
5. Rolc, S., Křesťan, J., Kopkáně, D., Štoller, J., Maňas, P.: Research of prospective materials for military protective structures. Appl. Mech. Mater. **796**, 35–42 (2015). ISSN 1660-9336
6. Figuli, L., Magura, M., Kavický, V., Jangl, Š.: Application of recyclable materials for an increase in building safety against the explosion of an improvised explosive device. Adv. Mater. Res. 1001, 447–452 (2014). Trans Tech Publications, Switzerland 2014. doi:10.4028/www.scientific.net/AMR.1001.447

7. Štoller, J., Dvořák, P.: Cementitious composite materials for critical infrastructure. In: Proceedings of 20th International Scientific Conference Transport Means 2016. Kaunas University of Technology, Kaunas, Lithuania, pp. 9–10 (2016). ISSN 1822-296X
8. Rejmont, J., Štoller, J.: Design of building protection in peacekeeping operations of the armed forces of the Czech Republic based on simulations of the effects of blast wave. In: 2015 International Conference on Military Technologies ICMT 2015, art. no. 7153674 (2015). ISBN: 978-807231976-3. doi:10.1109/MILTECHS.2015.7153674
9. Krátký, J., Trtík, K., Vodička, J.: The steel fibre reinforced concrete structures, 108 pages. Information Center of ČKAIT, Prague (1999). ISBN 80-86364-00-3
10. Jianming, G., Wie, S., Keiji, M.: Mechanical properties of steel fiber-reinforced, high-strength, lightweight concrete. Cem. Concr. Compos. **19**(4), 307–313 (1997). ISSN 0958-9465, http://dx.doi.org/10.1016/S0958-9465(97)00023-1
11. Suaris, W., Shah, S.: Properties of concrete subjected to impact. J. Struct. Eng. **109**(7), 1727–1741 (1983). doi:10.1061/(ASCE)0733-9445(1983)109:7(1727)
12. Štoller, J., Dvořák, P.: Non-destructive testing of full-scale slabs before and after blast load. In: 2015 Proceedings of the International Conference Transport Means, pp. 298–301 (2015). ISSN: 1822296X
13. Czech technical standard ČSN ISO 1920-10. Testing of concrete - Part 10: Determination of static modulus of elasticity in compression, 12 pages. ÚNM Publishing, Prague (2016)
14. Information on. http://www.betonracio.sk/betonracio/downloads/modul_pruznosti.pdf
15. Information on. http://stavba.tzb-info.cz/beton-malty-omitky/10746-dynamicke-a-staticke-moduly-pruznosti-betonu
16. Czech technical standard ČSN 73 1319. Testing of concrete - Part 10: Determination of static modulus of elasticity in compression, 12 pages. ÚNM Publishing, Prague (2016)
17. Czech technical standard ČSN 73 1371. Ultrasound pulse method of concrete testing, 20 pages. ÚNM Publishing, Prague (1982)

Design and Assessment of Shape of Protective Structure by Usage of CFD Software Environment Ansys Fluent

Jiří Štoller(✉) and Branislav Dubec

University of Defence, Kounicova 65, 662 10 Brno, Czech Republic
{jiri.stoller, branislav.dubec}@unob.cz

Abstract. The purpose of the study was to design and assess the shape of the construction of the protective structure with regard to its building materials and technical possibilities for its production. Another indispensable limiting condition was the requirement for its easy transportability, assembly and modularity. The optimal shape solution was sought out by using CFD simulation carried out in the software environment of ANSYS Fluent. Appropriateness of different variants of the protective structure, differing form each other by parameterized shape, was assessed on the basis of a drag coefficient. After obtaining a representative number of measured values of the frag coefficient for each variant, results were processed using optimization methods built in the software environment Ansys Workbench, which has found its minimum value. On the basis of this value the most suitable candidate for the optimal variant of the shape of protective structure was chosen. This candidate was subsequently evaluated using the simulation for verification of forecasted value of its drag coefficient.

Keywords: Protective structures · Shape optimization · CDF · Ansys Fluent · Drag coefficient

1 Introduction

The activity of the Army of the Czech Republic abroad are now largely confined to operation in peacekeeping, stabilization or observation missions. Integral part their action is the construction of facilities in the form of bases which allow them to fulfil their tasks in dangerous and risky environment. The security of the base and its protective qualities are result of a series of measures, where building protective structures is their integral part. The ability of force protection belongs among the most important properties of protective structures, where the level of protection is evaluated in terms of both, the resistance against pressure in the forefront of the blast wave and resistance to conventional means of destruction [7]. Military bases and camps are attacked from the surrounding terrain or urban areas using unguided missiles, mortars and anti-tank weapons, as well as are attack using improvised explosive devices carried by persons, luggage or vehicles. The last mentioned methods of warfare, based on experience from foreign missions of NATO armies, are coming to the forefront as well as protection against them because of theirs unpredictability and effectiveness [8].

A. Kravcov et al. (eds.), *Durability of Critical Infrastructure, Monitoring and Testing*, Lecture Notes in Mechanical Engineering, DOI 10.1007/978-981-10-3247-9_23

Up to the present day, protective structures in the ACR are designed using the technical manual Zen 2-1/1 [6] and its complements as well as the publications Methodology for designing ACR bases in foreign missions [5]. Constructions of protective structures built according these publications are not primarily focused on their resistance to the effects of the blast waves and do not meet one or more demands placed on contemporary protective structures, such as simplicity of assembly, variety of transportation options, modularity, low weight of the construction or low demands on machinery, equipment or human resources.

The need of possibility of fast and easy construction of protective structures with sufficient protective attributes for the need of both, the army personnel and the civilian members accompanying them into their risk environment of action, sought to design a new protective structure.

2 CAD Model and Its Parametrization

Before the creation of the model in CAD software it was necessary to take into account several restrictive conditions, which were imposed on the shape and dimensional arrangement of the model, based on which the actual structure will be finalized. One of the first conditions was associated with the manufacturing capabilities. The structure designed in this study will have manufactured at a reasonable cost, conventional manufacturing procedures and from materials based on concrete. Another requirement was the ability to assemble the building from number of major parts. This will ensure its modularity and will ensure variety of options for its transportation. The requirement for placement of the entrance door frame in a plane perpendicular to the ground as well as the minimum requirements for a dimensional arrangement of the shelter further limits the possibilities of shape solution of the future construction.

In addition to the restrictions relating to the structure itself, the draft of the building also had to take into account methods used in the process of the search, evaluation and optimization of the shape of the building. The relative long durations of the CFD simulations forces to exploit symmetries and simplifications of the geometry of the studied objects. To create a CAD model Autodesk Inventor 2015 was chosen due to its compatibility with Ansys Workbench software environment and its components.

Based on all requirements and conditions mentioned in previous paragraphs, it was decided to start the finding of the shape of the new protective structure from the cuboid, most commonly used shape of protective structures. Since a similar kind of structures can be assembled from individual concrete frames, a similar concept can be utilized for the construction of the new structure. Due to technical and manufacturing restrictions, rotationally symmetrical shapes with an infinite number of symmetry planes were discarded. However, it is appropriate to keep on the model at least two planes of symmetry, where one plane also serves as a symmetry planes for dividing the model into half for lowering the computing time demands of the simulations. For the same reasons the models were created in the scale of 1:10.

Figure 1 shows the YZ plane of symmetry of the model. To reduce the computational complexity of tasks, only one half of the model bounded by this half-plane will be used in the simulations. Figure 2 shows a plane of symmetry YX, which represent a plane of symmetry established on the basis of requirements to simplify manufacturing processes of the real construction. To further simplify the computing model, reducing the number of parameters and requirement of simplicity of manufacturing processes, the sidewalls of model stayed without change from the basic model, i.e. without changing its curvature in the plane of symmetry YX (Figs. 1 and 2).

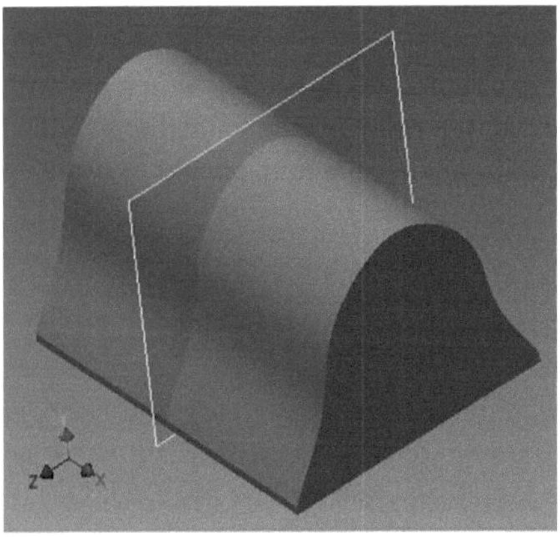

Fig. 1. Plane of symmetry YZ

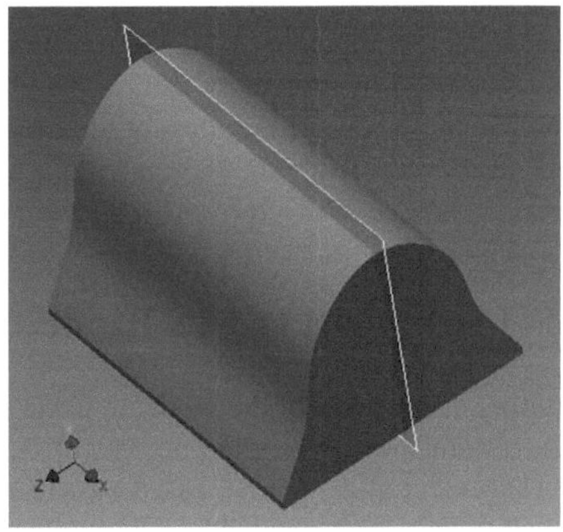

Fig. 2. Plane of symmetry YX

For the purpose of the simulation, constant and parameterized dimensions of the model were chosen. The change in the value of parametrized dimensions provides change in the overall geometry of the model in the main walls. For parameterized dimensions were chosen the distances of points 1 to 5 labeled DS_1 to DS_5 (marked red) from the Y axis (Fig. 3), where their interpolation curve combined witch two auxiliary points (marked green) creates the geometry of the surface of the structure.

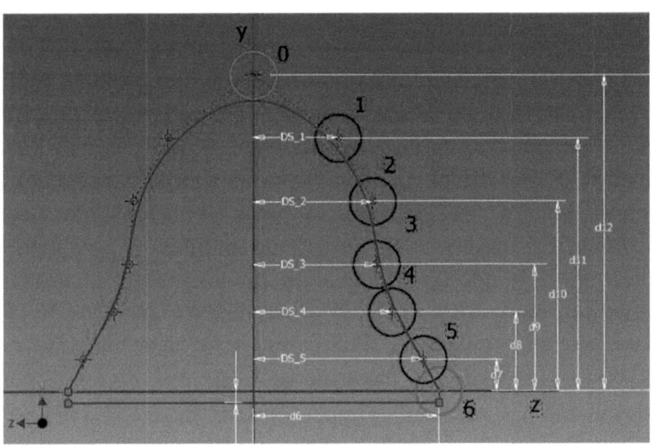

Fig. 3. Parametrized points placement and their interpolation curve (Color figure online)

By specifying the appropriate values and ranges of this parameters and by the usage of the interpolation curve, it is possible to obtain a wide variety of model shapes (Fig. 4). This method of forming the shape also ensured that all models of the protective structures have the same height and unchanged width at the base.

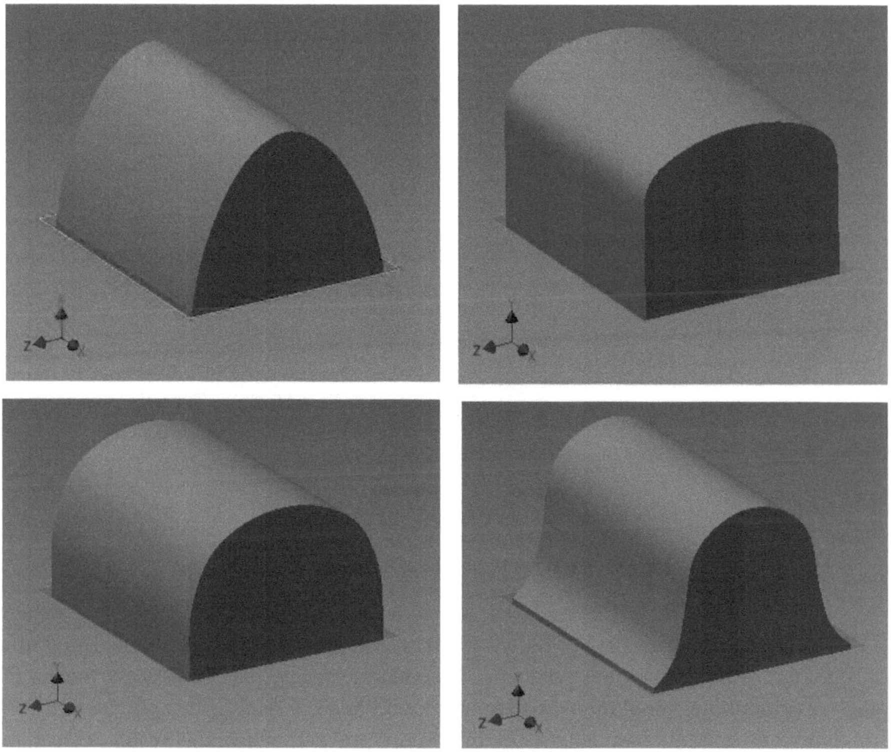

Fig. 4. Examples of the model shapes created by the usage of the parametrization

3 CFD Simulation

The reason for using CFD simulation as a starting method of dealing with the main objective of this paper was to determine the aerodynamic characteristics of the designed construction models. CFD simulation represent the first step in the process of finding the shape and overall design of the new protective structure. Comparison of the aerodynamic quality of the models with the results from assessment by other methods will follow, such as acceleration values of the models under the blast wave load [3] obtained from the simulation by the usage Ansys Autodyn software environment.

Ansys Fluent software environment dealing with computer simulation of the flow. The program uses for its calculations the numerical finite volume method, which solves the equations of conservation of mass and momentum. For tasks containing heat transfer and compressibility, equations describing the law of conservation of energy are added [1].

Preparing CFD simulation begins with creating geometry. Prepared CAD models was loaded into the component Design Modeler (DM) within the ANSYS Workbench environment. Subsequently, the computational domain was created, representing the area where the medium will flow (Fig. 5). For simplifying the simulation, in this step two planes of symmetry as mention before (Figs. 1 and 2) were created.

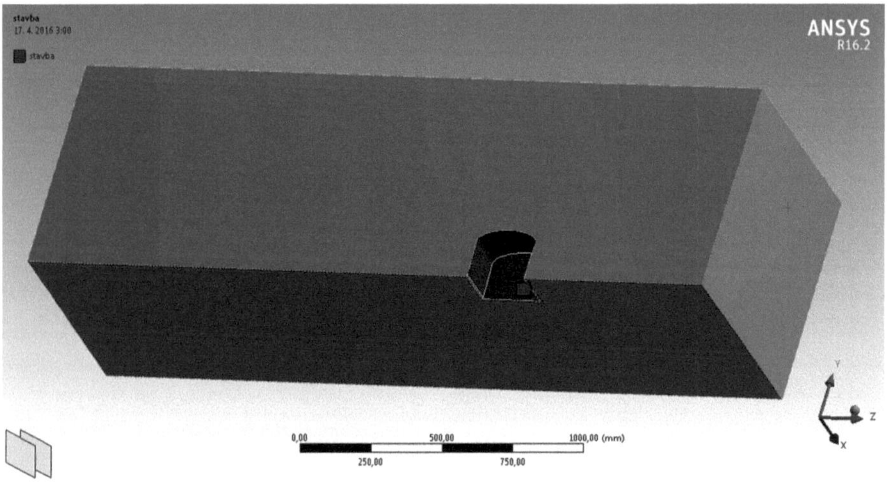

Fig. 5. Computational domain of the simulation

Creating a numerical grid of finite volume elements (meshing) is the next step after the creation of computational geometry. It constitutes a second preparatory step in the process of creating simulation in Fluent (FluentFlow). Creating the grid needs CFD simulation is difficult and extremely important process, since the result affects the amount of other processes and parameters, such as speed and accuracy of the calculation. Figure 6 shows the final version of mesh used in the simulations.

Afterwards, model with a grid was imported into the FLUENT, where the simulation was taken place. Before its initialization, it was necessary to choose the initial settings and parameters of the simulation through the tree of options allowing access to the settings of the element or part of the domain.

One of the most important setting option is located in the tree structure under the named General and contains configuration settings to grid controls and selection of the appropriate solver. In our case, Pressure - based solver was selected, as our flow was considered incompressible. The solver was also selected from the reason, that in this step of the process of finding the optimal shape solution, our aim is not to simulate the propagation of the blast wave, but to capture the magnitude of change of the output parameter – drag coefficient to subtle changes in the geometry of the shape of construction under the flow velocities in the magnitude of tens m/s. This simulation model has enabled us to perform a large number of repetitions of the simulation because of the favorable computational time demands.

Fig. 6. Final mesh version

The second most important control settings are connected with boundary conditions. Boundary conditions determine how the computational model behave during simulation in the frontier areas of the computational domain. Condition Velocity - Inlet creates the area of the input of the medium into the computational space. The value of the medium velocity entering through this wall was set to 70 m/s and the direction has been set in the direction of the negative z axis.

Before the optimization of the shape there was a need to verify the successful simulation settings, convergence of the solution and verification of the capabilities of the computational model to capture the change of shape of the designed models. To assess the results of the simulation stand-alone component CFX post were used, a component which is able to be linked with Fluent, acquire its outputs and subsequently this outputs graphically and numerically display.

Two variants of the models were assed, which differed by the value of its five input parameters. The third assessed model was separately designed in the shape of cuboid, which represents the shape of protective structures most commonly used. The parameter values as well as values of the output parameter displays Table 1. For the output parameters the drag coefficient was chosen.

Table 1. Model parameter values

Model	Parameter DS_1 [mm]	Parameter DS_2 [mm]	Parameter DS_3 [mm]	Parameter DS_4 [mm]	Parameter DS_5 [mm]	Force −z [N]	Drag coefficient C_d
A	150	150	150	150	150	45,53	0,53
B	80	110	130	140	147	47,86	0,56
C	Cuboid, $A \times B$, $A = 240\,\text{mm}$, $B = 190,7\,\text{mm}$					110,08	1,282

When comparing the graphical representations of the results of the simulation in the form of velocity contours, clear differences in velocity distribution can be seen. In Fig. 8 earlier flow separation and greater wake is present compared with Fig. 7. Figure 9 shows that the attachment of the flow are not present, which corresponds with large wake and the highest value of the drag coefficient C_d.

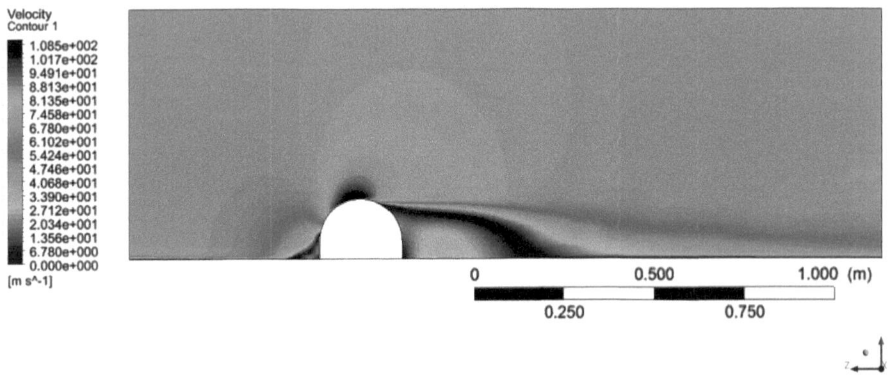

Fig. 7. Velocity contours – Model A

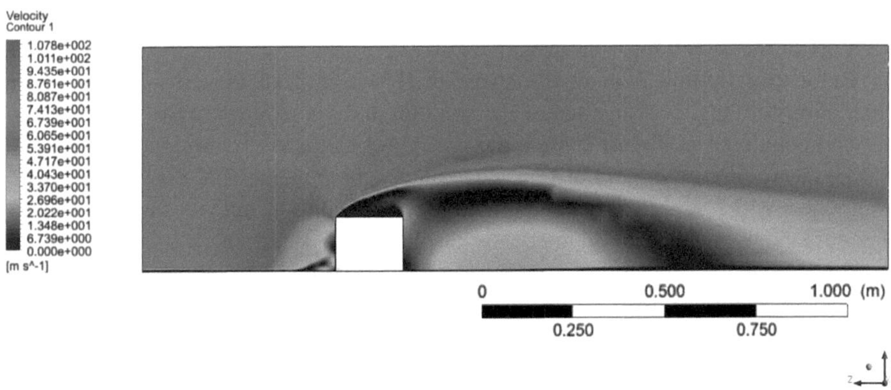

Fig. 8. Velocity contours – Model B

Based on the progress and results of simulations it is possible to conclude that convergent solution of the computational model was found, and also that the model is sufficiently accurate, and therefore useful for capturing relatively small changes in shape of the input CAD model variants of the protective structure.

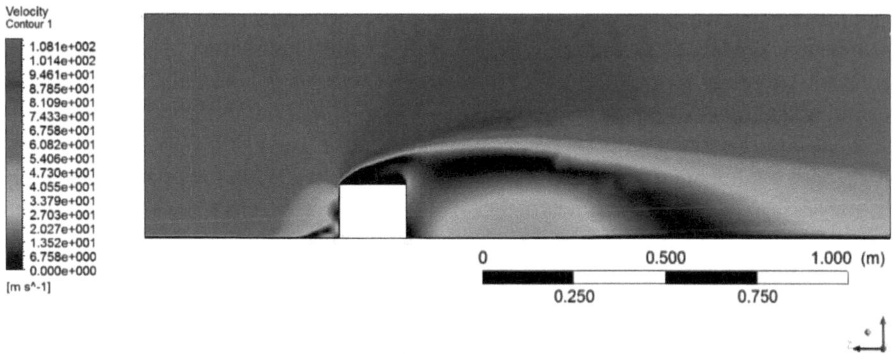

Fig. 9. Velocity contours – Model C

4 Optimization

Optimization is an integral part of any process of developing a final design or engineering problem solving. General description of such process can be divided into six steps (Fig. 10) related to the preparatory step in which a parameterization of input values and creation of a computational model of the problem is made. During the first step a few calculations from manually inserted values of input parameters based on expert estimates are done (What if study), and subsequently on their bases sensitivity studies and correlation (Correlation Study) are conducted on these input parameters. This will help reduce their numbers on the most important, most affecting the solution and thereby reduce the number of calculations needed during the next step to establish a representative number of alternative solutions (Design of Experiment, DOE). On their basis the entire solution space is created (Response surface), which can using mathematical models predict the value of the output parameters based on the input parameters throughout its defined range. The actual optimization (Optimisation) follows, in which optimal solution with regard to the desired goal is found. The final step in solving engineering problems is to implement Six Sigma analysis, which ensures that the results of the calculation model will be feasible and robust in practice [2].

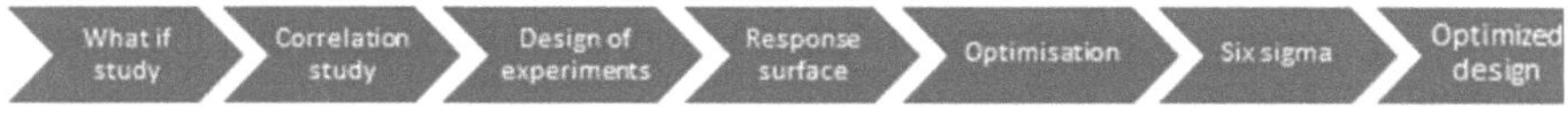

Fig. 10. General optimization process [2]

As a tool for optimization the software component DesignExplorer integrated into the software environment ANSYS Workbench was chosen.

In previous section of this paper the creation of the CAD model, its parametrization, the creation of simulation and its results for three sets of input parameter values was described. It was verified that the solution converges and computing model is accurate enough to capture the defined geometry changes. Since the change of any of the five input parameters causes a discernible change in the geometry of the model and output parameters, neither of them was expelled from further assessment. Therefore, it was proceeded directly to creation of a representative sample solution (DOE). DOE consisting of a number of representative variations of the input parameters forming a single body design (Design Point, DP) and corresponding output parameters. DOE creation starts with the selection of input parameters, which are needed to include into the study. In our case, five input parameters DS_1 to DS_5 were selected. And two output parameters, the drag coefficient C_d, respectively force acting in the negative z axis were selected. Each input parameter was defined by its continuous range of (Table 2).

Table 2. Input parameter values and their range

Parameter name	Range [mm]
DS_1	60–120
DS_2	70–120
DS_3	75–140
DS_4	80–120
DS_5	90–120

In the next step the Response Surface was created. Its advantage is that it provides approximated values of output parameters based on input parameters anywhere in the analyzed space without the need to perform calculation for all these input values. It is created on the bases of the outputs of the previous step of the optimisation process, i.e. the values provided by the DOE.

Because by using the RS we are able to describe the dependence of the output parameter C_d on the input parameters, an optimization goal was to find the design point of the minimal value of its output parameter. For this purpose the recommended method MISQP [2] were used. Since the value of drag coefficient candidates found in this optimization step has the nature of predictions, the predicted output value was subsequently verified by CFD simulations. Thus obtained points were used again to create higher quality RS with their inclusion among design points, which helped to increase the accuracy of prediction and probability of finding the actual design point with minimal output parameter.

Table 3 shows model with its input and minimal output parameters found in the process of optimization and five other selected models ordered by their output parameters.

On Fig. 11 the shape of the models with the lowest and highest value of drag coefficient from the Table 3 can be seen.

5 Summary

This work presents the first step in design of new protective structure to the effects of the explosive blast. As the first method of finding the most appropriate shape of protective structures was chosen CFD simulation, performed in the software

Table 3. Input parameter values and its range

Model number	Parameter DS_1 [mm]	Parameter DS_2 [mm]	Parameter DS_3 [mm]	Parameter DS_4 [mm]	Parameter DS_5 [mm]	Force –z [N]	Drag coeff C_d
1.	88,3	119,4	112,6	95,1	112,5	44,67	0,5186
2.	83,5	101	105	107	110,7	47,12	0,5471
3.	80	105	117	120	120	45,53	0,59
4.	60	73	91	100	112	55,61	0,645
5.	98	87,9	97,1	94,3	100,7	56,59	0,657
6.	Cuboid, $A \times B$, $A = 240\,\text{mm}$, $B = 190,7\,\text{mm}$					110,1	1,282

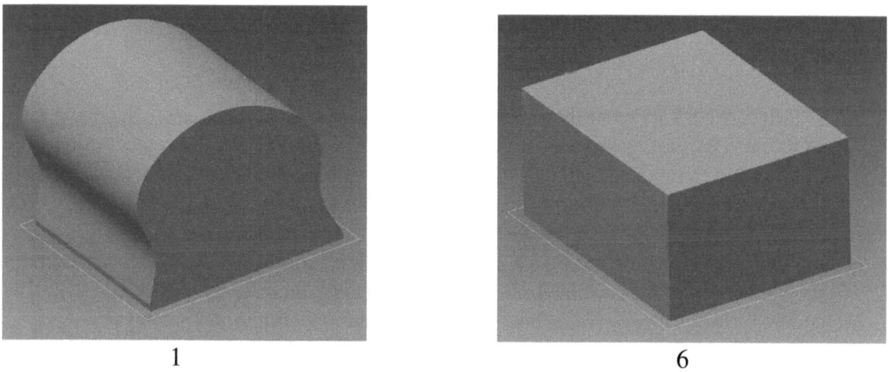

Fig. 11. Shapes of models no. 1 and 6 from Table 3

environment Ansys Fluent based on computing model of the external flow around the models of protective structure created with Autodesk Inventor 2015. A computational model that represents a compromise between accuracy and computational complexity managed to detect and describe the relatively small differences in the geometry of the individual shape variants of the assessed models. Using a set of optimization tools within the components integrated into the software environment ANSYS Workbench, an optimized solution was found, where the optimized criterion was represented by lowest drag coefficient. Besides the final shape, a set of other models are presented, due to theirs favourable shape form the manufacturing point of view. As a benchmark, the most basic shape of protective structure in the shape of cuboid is assessed.

The possibilities for further advancement in design of protective structures are extensive, such as the possibility of implementation of CFD simulations of flow around the construction under conditions closer to conditions of blast wave followed by optimization of its shape. Above all, the task is to increase the velocity of flow to values which excess of several times the speed of sound. Such conditions might be able to more accurately capture and predict the principles of flow of the blast waves and thus the structure optimized in such circumstances would provide a greater degree of resistance against the effect of blast waves and be capable to ensure the higher level of force protection. The next step in creation of the new protective structure should be the comparison of the aerodynamic quality results from assessment of the models under the blast wave load obtained from the simulation by the usage Ansys Autodyn software environment, explicit analysis tool for modeling the nonlinear dynamics of solids, fluids, gases and their interactions [4]. This comparison will provide the information how effective was the CFD method in finding the optimized shape of the protective structure.

Acknowledgment. The work presented in this article has been supported by the Czech Republic Ministry of Defence - K 201 Department development program "Development of technologies in weapons construction, ammunition, weapons instrumentation, materials engineering and military infrastructure".

References

1. ANSYS, Inc. ANSYS Fluent theory guide, Release 15, USA, SAS IP, Inc., November 2013
2. ANSYS, Inc. Design Exploration, Release 12.1, USA, SAS IP, Inc., November 2009
3. Boyd, S.D.: Acceleration of a plate subject to explosive blast loading - trial results. Australia, Department of Defence, Maritime Platforms Division Aeronautical and Maritime Research Laboratory, March 2000
4. Coufal, D.: Modelling of effects of blast wave on the protective structures. Dissertation Thesis. University of defence in Brno, Faculty of military technology, Brno, 102 p (2013)
5. Methodology for designing ACR bases in foreign missions. http://ipzv.unob.cz/
6. Field fortification: Surface and soil-covered shelters. Zen-2-1/1 Prague: MNO (1972)
7. STANAG 2280 - MC ENGR (Edition 1) (Ratification draft 1) – Design Threat Levels and Handover Procedures for Temporary Protective Structures. NATO Standardization Agency, June 2007
8. Zalesky, J.: Military engineering support of force protection in foreign missions. Dissertation thesis. University of defence in Brno, Faculty of economy, Brno, 162 p (2012)

The Basic Properties of Materials Suitable for Protective Structures and Critical Infrastructure

Jiří Štoller(✉) and Eva Zezulová

University of Defence Brno, FVT, Kounicova 65, 66210 Brno, Czech Republic
{jiri.stoller, eva.zezulova}@unob.cz

Abstract. The deals with the basic properties of cement-based materials that are appropriate for protective structures and critical infrastructure. The basic properties of materials were tested on the material samples. The samples were tested before loading by the distant explosion, by the penetration and by the puncture. During the material testing were measured the compressive strength, the tensile strength and the dynamic modulus of elasticity. The article discusses only the compressive strength and the tensile strength.

Keywords: Compressive strength · Tensile strength · Toughness

1 Introduction

The worsening of the security situation in Europe and over the world leads to the need for enhanced protection of critical infrastructure facilities. At the same time the unfavorable security situation contributes to development of new constructions and new materials military shelters. And that is why the materials which are commonly used for protective shelter building and for object of critical infrastructure building are not sufficiently resistant to new weapons. Protection of objects of critical infrastructure is crucial especially at this time when there is an increased incidence of terrorist attacks.

Development of new materials for protective structures that would suit better to the new demands of safety circumstances and they better withstand the loads, they may contribute to reducing of the extent of the damage caused by e.g. a distant explosion, penetration and puncture. Buildings or structures that can be subjected to a bomb attack by a terrorist organization or offenders are considered as one of the critical infrastructure elements [1].

One of the materials that could satisfy such demanding requirements is steel fibre concrete. Its properties make it ideally suited to the category of materials that could withstand the extreme loads. Currently steel fibre reinforced concrete mostly used as a material for construction of industrial floors [2]. Especially because of these constructions is steel fibre reinforced concrete has become a fixed part of our range of construction materials. This use, however, is far from exhaust the wide application of steel fiber reinforced concrete construction. Even in the Czech Republic are already a number of examples of structures during their construction the steel fibre reinforced concrete has been used.

A. Kravcov et al. (eds.), *Durability of Critical Infrastructure, Monitoring and Testing*,
Lecture Notes in Mechanical Engineering, DOI 10.1007/978-981-10-3247-9_24

Moreover, for the production of steel fiber reinforced concrete can be used recycled material in the form of wires, which were obtained from used car tires. Application of recyclable materials for an increase in building safety against the explosion of an improvised explosive device described in [3].

Examples of possible use of steel fiber reinforced concrete are contained in Directive [4]. Mastering the technology and knowledge of the properties of steel fiber reinforced concrete is a guarantee that this promising material will be used more often in the construction. The relatively high price of wires in the Czech Republic is an unpleasant circumstance that prevents the expansion of the use of steel fiber reinforced concrete.

2 The Compressive Strength

Strength tests of the fibre reinforced concrete and the plain concrete specimens were conducted in the laboratory of the Institute of Building Testing VUT Brno. Test of the compressive strength was conducted in accordance with the Czech technical standard ČSN EN 12390-3 [5] and the tensile bending strength was in accordance with the Czech technical standard ČSN EN 12390-5 [6].

The compressive strength was tested on cubes. A total amount of 20 pieces of concrete testing cubes and 18 pieces of fibre reinforced concrete testing cubes. The manufactured testing cube have length of the edges 150 mm. The cubes were made and treatment according to the Czech technical standard ČSN EN 12390-2 [7].

Before destructive testing was measured the density of hardened concrete on cubes according to the ČSN EN 12390-7 [8]. The method of the density calculating from the measured actual dimensions has being used to determine the density. Each cube was testing with using the ultrasound device TICO. Ultrasonic testing was performed and evaluated according to the Czech technical standard ČSN 73 1371 [9]. The evaluation of this test is shown in [10]. After the non-destructive ultrasonic testing method was on each cube destructively determined compressive strength (see Fig. 1). Measured values of concrete strengths for the individual test samples are presented in Table 1. Measured values of steel fibre reinforced concrete strengths are presented in Table 2. Normal distribution of compressive strength of concrete is presented in the Graph 1. Normal distribution of compressive strength of steel fibre reinforced concrete is presented in the Graph 2.

Calculation of compressive strength was carried out according to the Czech technical standard ČSN EN 12390-3 [5].

$$f_{c,cube} = \frac{F}{A_c} \tag{1}$$

where:

$f_{c,cube}$ compressive strength [MPa],

F maximum load at crushing [N],

A_c loaded surface of the test specimen [mm^2]

Fig. 1. Compressive strength testing - testing machine HECKERT 300 t

The evaluation of a minimum characteristic cube compressive strength was carried out according to Czech technical standard ČSN 73 0038 [11] and ČSN 73 20 11 [12].

$$R_{bg} = R_{b,prům,(m)} - \beta_n \cdot s \equiv f_{ck,cube} \equiv R_{fd} \quad (2)$$

where:

$R_{b,prům,(m)}$ median of the concrete compressive strength,

β_n coefficient of estimation 5 % of the quantity (for 20 pcs it is 1.71 and for 18 pcs it is 1.722 [11]),

s sample standard deviation

Compressive strength of unreinforced concrete

$$R_{bg} = R_{b,prům,(m)} - \beta_n \cdot s \equiv f_{ck,cube}$$

$$f_{ck,cube} = 113.09 - 1.71 * 12.2145 = 92.20\,\text{MPa}$$

→ strength class of concrete **C 70/85**

The steel fibre reinforced concrete compressive strength

$$R_{bg} = R_{b,prům,(m)} - \beta_n \cdot s \equiv R_{fg}$$

Table 1. The measured values chart of the concrete compressive strength

I.D.	Sample dimensions [mm]			Weight	Sample volume	Density	The crushing load	Compressive strength $f_{c,cube}$
	a	b	c	[kg]	m^3	kg/m^3	[kN]	$[N/mm^2]$
1	148.0	147.5	150.0	8.5719	0.003275	2618	2780	127.3
2	149.0	149.5	149.8	8.5862	0.003337	2573	2313	103.8
3	149.8	148.2	149.6	8.5642	0.003321	2579	2028	91.3
4	149.6	149.6	149.7	8.5854	0.003350	2563	2052	91.7
5	149.7	148.8	149.7	8.5704	0.003335	2570	2432	109.2
6	150.2	149.8	149.6	8.6886	0.003366	2581	2749	122.2
7	147.8	149.5	149.6	8.6364	0.003306	2613	2487	112.6
8	149.8	150.0	149.2	8.6032	0.003353	2566	2838	126.3
9	149.7	149.6	149.4	8.4998	0.003346	2540	2673	119.4
10	151.3	149.5	149.6	8.6660	0.003384	2561	2135	94.4
11	150.6	149.8	149.9	8.5758	0.003382	2536	2542	112.7
12	149.7	150.0	151.0	8.6465	0.003391	2550	2780	123.8
13	150.2	150.1	150.2	8.6520	0.003386	2555	2808	124.6
14	149.8	150.2	149.7	8.6206	0.003368	2559	2878	127.9
15	150.1	149.8	149.9	8.5681	0.003370	2542	2241	99.7
16	149.4	151.2	149.5	8.6538	0.003377	2562	2938	130.1
17	152.2	149.6	149.8	8.7420	0.003411	2563	2732	120.0
18	149.5	149.6	149.7	8.6181	0.003348	2574	2437	109.0
19	149.5	149.6	149.7	8.7275	0.003348	2607	2577	115.2
20	149.6	149.7	150.2	8.6240	0.003364	2564	2578	115.1

Table 2. The measured values chart of the compressive strength of the steel fiber reinforced concrete

I.D.	Sample dimensions [mm]			Weight	Sample volume	Density	The crushing load	Compressive strength R_f
	a	b	c	[kg]	m^3	kg/m^3	[kN]	$[N/mm^2]$
22	149.3	150.2	149.8	8.7195	0.003359	2596	2572	114.7
23	150.0	149.8	151.3	8.6546	0.003400	2546	2639	117.4
24	149.6	150.3	150.0	8.5973	0.003373	2549	2415	107.4
25	149.5	150.3	159.6	8.5729	0.003586	2391	2203	98.0
26	150.0	150.0	151.8	8.6210	0.003416	2524	2440	108.4
27	149.2	149.5	151.6	8.6097	0.003381	2546	2450	109.8
28	149.4	149.7	150.1	8.6121	0.003357	2565	2550	114.0
29	149.9	149.8	150.2	8.6542	0.003373	2566	2640	117.6
30	150.1	150.2	149.7	8.6970	0.003375	2577	2630	116.7
31	149.6	149.7	149.9	8.5982	0.003357	2561	2460	109.8
32	149.9	150.0	150.3	8.6734	0.003379	2566	2350	104.5
33	150.2	150.1	150.1	8.7134	0.003384	2575	2560	113.6
34	149.7	149.8	150.1	8.6901	0.003366	2582	2450	109.3
35	149.8	149.7	150.1	8.7200	0.003366	2591	2640	117.7
36	149.8	149.6	150.1	8.6542	0.003364	2573	2530	112.9
37	149.7	149.9	150.2	8.6790	0.003370	2575	2470	110.1
38	149.6	149.8	150.1	8.6902	0.003364	2583	2480	110.7
39	150.1	150.0	150.2	8.7200	0.003382	2579	2620	116.4

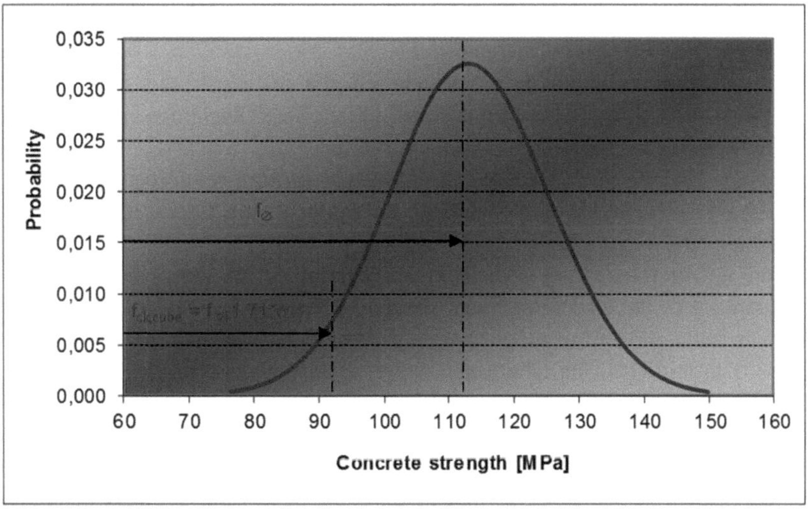

Graph 1. The normal distribution of the compressive strength of concrete N(113,09;12,21)

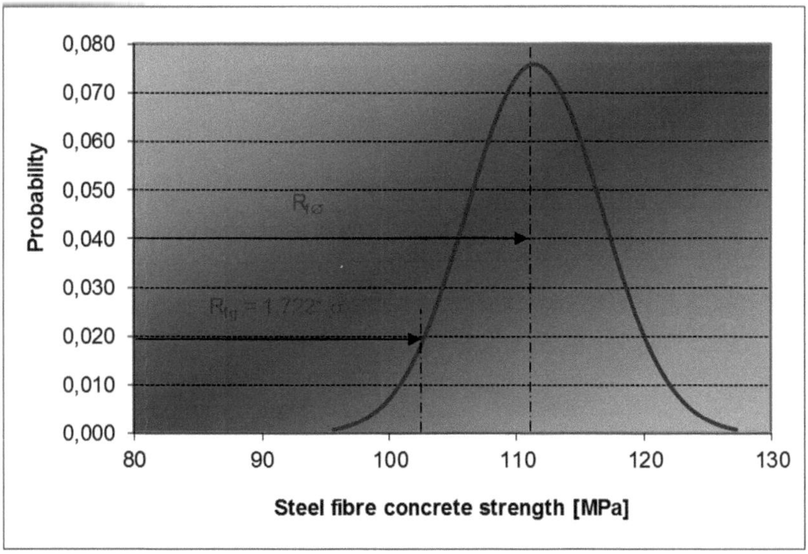

Graph 2. The normal distribution of the compressive strength of steel fibre reinforced concrete N(111,43;5,26)

$$R_{fg} = 111.4296 - 1.722 * 5.2612 = 102.37\,\text{MPa}$$

$\Rightarrow$ strength class of steel fibre reinforced concrete **C 80/95**

3 Tensile of Specimens for the Flexural Strength Testing

For the test of flexural strength was chosen the testing beams with dimensions 150 × 150 × 600 mm. These dimensions of the test specimen were determined especially with regard to the type and length of used steel wires [4]. A total amount of 20 pieces of concrete testing beams and 18 pieces of fibre reinforced concrete testing beams. Beams were made and treatment according to the Czech technical standard ČSN EN 12390-2 [7].

Before destructive testing of beams was determined the density of hardened concrete according to the Czech technical standard ČSN EN 12390-7 [8]. The density of beams was calculating from the measured actual dimensions and weight of the beams with steady humidity in the laboratory.

Each beam when testing was stressed by two burdens which were acting in thirds margin (see Figs. 2 and 3). Measured values of tensile strength of concrete beams are presented in Table 3. Measured values of tensile strength steel fibre reinforced concrete beams are shown in Table 4. Normal distribution of flexural strength of concrete is illustrated in Graph 3. Normal distribution of flexural strength of steel fiber reinforced concrete is illustrated in Graph 4.

Fig. 2. Testing of specimens for flexural strength

The evaluation of tensile strength in bending was done according to the Czech technical standard ČSN 12390-5 [6].

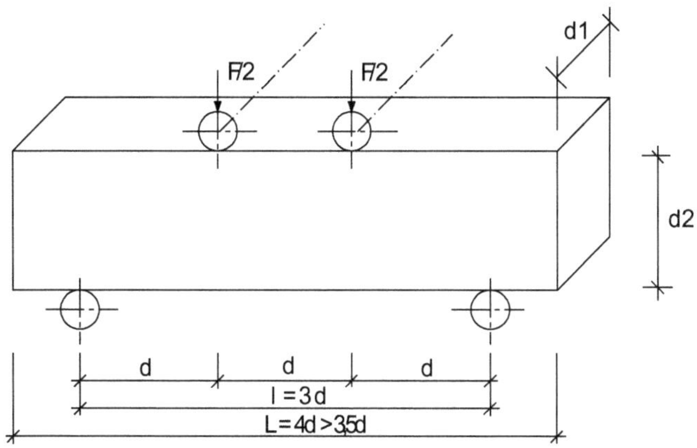

Fig. 3. The arrangement of forces in loading of the test specimen

Table 3. Table of flexural strength measured on concrete beams

I. D.	Sample dimensions [mm]			Weight	Sample volume	Density	The crushing burden	Tensile strength f_{ct}
	a	b	c	[kg]	m^3	kg/m^3	[kN]	$[N/mm^2]$
1	151.1	152.3	596.2	34.6830	0.013720	2528	53	7.6
2	151.0	150.2	597.3	34.6230	0.013547	2556	67	9.8
3	151.2	151.3	596.3	34.6420	0.013641	2539	58	8.4
4	150.2	151.3	597.8	34.6820	0.013585	2553	58	8.4
5	150.2	151.5	598.6	34.6750	0.013621	2546	49	7.1
6	151.3	152.2	597.8	34.7820	0.013766	2527	63	9.0
7	151.2	152.1	596.2	34.6850	0.013711	2530	68	9.7
8	151.2	152.2	597.6	34.6920	0.013752	2523	52	7.4
9	151.3	152.3	596.4	34.7850	0.013743	2531	47	6.7
10	152.3	151.3	598.2	34.7560	0.013784	2521	56	8.0
11	151.2	151.3	598.6	34.7820	0.013694	2540	49	7.1
12	151.3	151.0	597.4	34.6930	0.013648	2542	68	9.9
13	151.2	152.3	598.2	34.8100	0.013775	2527	54	7.7
14	151.2	152.0	597.6	34.7950	0.013734	2533	58	8.3
15	151.2	151.3	598.4	34.8120	0.013689	2543	53	7.7
16	151.2	152.4	597.3	34.8210	0.013764	2530	52	7.4
17	151.2	152.1	596.5	34.7654	0.013718	2534	48	6.9
18	151.2	152.1	597.4	34.7562	0.013739	2530	58	8.3
19	151.3	151.2	597.6	34.7850	0.013671	2544	62	9.0
20	151.4	151.2	597.6	34.8300	0.013680	2546	55	7.9

Table 4. Table of flexural strength of steel fibre reinforced concrete beams during its loading

I. D.	Sample dimensions [mm]			Weight	Sample volume	Density	The crushing burden	Compressive strength R_{ft}
	a	b	c	[kg]	m^3	kg/m^3	[kN]	[N/mm^2]
22	151.2	150.3	598.3	34.8240	0.013597	2561	62	9.1
23	151.3	151.2	597.6	34.8120	0.013671	2546	66	9.5
24	151.2	152.4	597.3	34.7850	0.013764	2527	63	9.0
25	151.2	152.1	597.6	34.8410	0.013743	2535	58	8.3
26	151.1	151.4	597.2	34.6780	0.013662	2538	57	8.2
27	151.3	151.4	597.3	34.7630	0.013682	2541	56	8.1
28	151.2	152.4	598.2	34.8210	0.013784	2526	63	9.0
29	151.2	151.2	597.6	34.8130	0.013662	2548	63	9.1
30	151.3	151.2	597.5	34.7720	0.013669	2544	65	9.4
31	151.2	151.3	597.8	34.7650	0.013676	2542	58	8.4
32	151.3	151.2	597.9	34.8130	0.013678	2545	62	9.0
33	151.1	151.3	598.2	34.6540	0.013676	2534	63	9.1
34	151.0	151.2	598.2	34.7850	0.013658	2547	60	8.7
35	151.3	152.1	597.8	34.7820	0.013757	2528	59	8.4
36	151.3	151.4	598.8	34.8240	0.013717	2539	63	9.1
37	151.4	151.3	598.7	34.8450	0.013714	2541	64	9.2
38	152.0	151.3	597.8	34.8820	0.013748	2537	62	8.9
39	151.5	152.1	598.7	34.8220	0.013796	2524	59	8.4

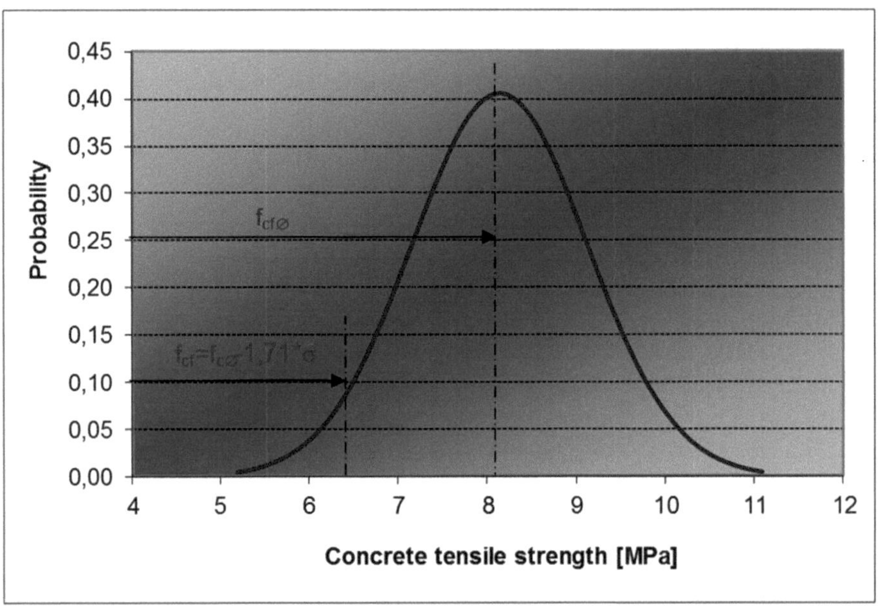

Graph 3. Normal distribution of the tensile strength of the plain concrete N(8,140;0,98)

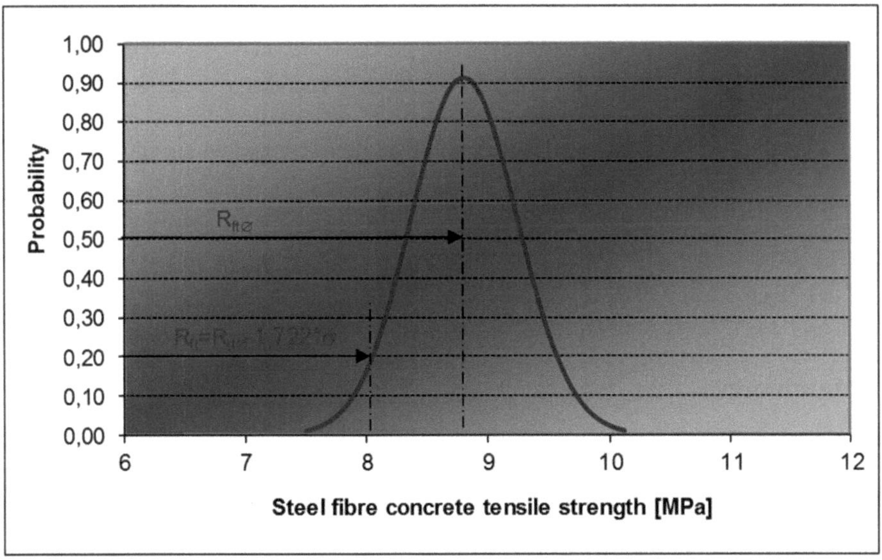

Graph 4. Normal distribution of the tensile strength for the steel fiber reinforced concrete N (8,812;0,44)

$$R_{bd} = R_{bd,\text{prům},(m)} - \beta_n \cdot s \equiv f_{cf} \equiv R_{ftg} \tag{3}$$

kde: $R_{b,\text{prům},(m)}$ median of concrete compressive strength,
β_n coefficient of estimation, 5 % of the quantity (for 20 pcs it is 1.71 and for 18 pcs it is 1.722 [11]),
s sample standard deviation

The evaluation of the minimum characteristic flexural strength was performed according to the Czech technical standard ČSN 73 0038 [11] and ČSN 73 20 11 [12].

$$f_{cf} = \frac{F \cdot l}{d_1 \cdot d_2^2} \tag{4}$$

where:
f_{cf} flexural strength [MPa],
F maximum load [N],
l the distance between the support rollers [mm],
d_1 a d_2 dimensions of the cross sectional view of the beam [mm]

Flexural strength of the concrete testing specimen

$$R_{bd} = R_{bd,prům,(m)} - \beta_n \cdot s \equiv f_{cf}$$

$$f_{cf} = 8.14 - 1.71 * 0.9816 = 6.46\,\text{MPa}$$

Flexural strength of the steel fibre reinforced concrete testing specimen

$$R_{ftd} = R_{ftd,prům,(m)} - \beta_n \cdot s \equiv R_{ftg}$$

$$R_{ftg} = 8.8112 - 1.722 * 0.4363 = 8.06\,\text{MPa}$$

4 Summary

For the testing of the concrete strength were made testing specimens in the shape of a cube with its edge 150 mm length. During this tests it was found that samples of normal concrete have a minimum compressive strength 91 MPa and a maximum compressive strength 130 MPa. After evaluation according to Czech technical standard ČSN 73 0038 [11] and ČSN 73 2011 [12], it was found that the minimum characteristic cube strength of the concrete it corresponds to the strength class of concrete C70 /85.

For steel fiber reinforced concrete was measured minimum compression strength 98 MPa and the maximal compressive strength 117 MPa. After evaluation according to the Czech technical standards ČSN 73 0038 [11] and ČSN 73 2011 [12] the samples were classified according to the minimum characteristic cube strength as the concrete which belongs to the strength class C80/95.

Flexural strength was tested on testing prisms that had a dimension 150 × 150 × 600 mm according to the Czech technical standard ČSN EN 12390-5 [6]. Evaluation of the minimum characteristic flexural strength was done according to the Czech technical standard ČSN 73 0038 [11] and ČSN 73 20 11 [12]. For samples made from the plain concrete was measured the minimum flexural strength 6.70 MPa, and the minimum flexural strength was measured 9.86 MPa. The minimum characteristic flexural strength of the plain concrete was 6.46 MPa.

The minimum flexural strength of the steel fiber reinforced concrete was measured 8.07 MP and maximal flexural strength was 9.54 MPa. The minimum characteristic strength of the steel fiber reinforced concrete flexural strength was 8.6 MPa.

By comparing these two materials we will find that mere addition of wires to the concrete structure, both measured parameters has increased. Both the compressive strength and the tensile strength in bending have increased. However, the important fact remains, that by the addition of wires into the concrete it causes steep increase of the dynamic resistance of the steel fiber reinforced concrete in comparison with plain concrete.

Acknowledgment. The work presented in this article has been supported by the Czech Republic Ministry of Defence - K 201 Department development program "Development of technologies in weapons construction, ammunition, weapons instrumentation, materials engineering and military infrastructure".

References

1. Dvořák, Z., Sventeková, E.: Evaluation of the resistance critical infrastructure in Slovak Republic. J. Eng. Manage. Competitiveness (JEMC) **3**(1), 1–5 (2013)
2. Trtík, K.: The steel fibre reinforced concrete in in the Czech Republic at the turn of the century. Construction sheets, No.12, pp. 22–23. ISSN 1211–4790 (1999)
3. Figuli, L., Magura, M., Kavický, V., Jangl, Š.: Application of recyclable materials for an increase in building safety against the explosion of an improvised explosive device. Adv. Mater. Res. 1001, 447–452. Trans Tech Publications, Switzerland (2014). doi:10.4028/www.scientific.net/AMR.1001.447
4. Krátký, J., Trtík, K., Vodička, J.: The Steel Fibre Reinforced Concrete Structures, 1st edn., 108 p. Information Center of ČKAIT, Prague (1999). ISBN 80-86364-00-3
5. Czech technical standard ČSN EN 12390-3. Testing hardened concrete - Part 3: Compressive strength of test specimens, 20 p. ÚNM Publishing, Prague (2009)
6. Czech technical standard ČSN EN 12390-5. Testing hardened concrete - Part 5: Flexural strength of test specimens, 12 p. ÚNM Publishing, Prague (2009)
7. Czech technical standard ČSN EN 12390-2. Testing hardened concrete - Part 2: Making and curing specimens for strength tests, 8 p. ÚNM Publishing, Prague (2009)
8. Czech technical standard ČSN EN 12390-7. Testing hardened concrete - Part 7: Density of hardened concrete, 12 p. ÚNM Publishing, Prague (2009)
9. Czech technical standard ČSN 73 1371. Ultrasound pulse method of concrete testing, 20 pages. ÚNM Publishing, Prague (1982)
10. Štoller, J., Dvořák, P.: Ultrasound diagnosis of protective structures after contact explosion. In: 2014 Proceedings of the 18th International Conference on Transport Means. Kaunas, Lithuania: Kaunas University of Technology, Lithuania, pp. 265–267 (2014). ISSN 2351-4604
11. Czech technical standard ČSN 73 0038. Assessment and verification of existing structures - Supplementary guidance, 56 pages. ÚNM Publishing, Prague (2014)
12. Czech technical standard ČSN 73 2011. Non-destructive testing of concrete structures, 23 pages. ÚNM Publishing, Prague (2012)

Laser-Ultrasonic Monitoring of the Gritical Structure Objects Produced from CRFC

Alexander A. Karabutov[1(✉)], Elena B. Cherepetskaya[1], Yulia G. Sokolovskaya[1], Elena A. Mironova[1], Dmitry V. Morozov[1], and Pavel Svoboda[2]

[1] Laser Ultrasonic Non-destructive Testing Lab, The National University of Science and Technology "MISiS", Leninsky Pr. 6, 119991 Moscow, Russia
aak@optoacoustic.ru,
{echerepetskaya,moroz_off_123}@mail.ru,
{julias-91,lenamironova71}@yandex.ru

[2] Department of Construction Technology, Faculty of Civil Engineering, Czech Technical University in Prague, Prague, Czech Republic
pavel.svoboda@fsv.cvut.cz

Abstract. Porosity is one of the most important composite structure failure, influenced on the properties of the construction and the residual service to fatigue. The non-destructive technologies able to detect the intimate structure transformation are well-come to the market. The comparative analysis of advantages and disadvantages of several technologies, such as X-ray tomography, conventional ultrasonic testing, thermal wave testing, remote laser-ultrasonic testing and contact laser-ultrasonic evaluation is presented. Contact laser-ultrasonic material evaluation seems to be the very technology for CFRF construction test. The advantages of contact laser-ultrasonic evaluation (CLUE) are described in detail. The possibility to evaluate the residual life-time of the objects of critical structure composite construction with CLUE is discussed.

Keywords: Laser-ultrasonic monitoring · Composite structure · Non-destructive testing · Delamination · Porosity · X-ray tomography

1 Introduction

Methods of non-destructive testing (NDT) are widely used for the assessment of carbon fiber composites. These methods are radiographic imaging techniques [1–3] including X-ray computed tomography [1–3], neutron tomography [1], Compton backscatter technique [3], thermal non-destructive testing [4, 5], acoustic emission and ultrasound techniques [6, 7]. But different modifications of X-ray methods have been widely used for metal inspection. Resolution of cracks is not as good with structural plastics as with metals because of the much lower density. Thermal non-destructive testing for inspection to detect defects in the carbon fiber reinforced composites is generally limited to a depth of 0.5 mm. Acoustic emission and ultrasound techniques [6, 7], including laser ultrasound, are more informative NDT methods for the investigation of structure and properties of CFRC.

Laser ultrasound provides pulsed acoustic beam with smooth phase front. The laser-excited ultrasonic pulse has smooth temporal shape without ringing. Described

A. Kravcov et al. (eds.), *Durability of Critical Infrastructure, Monitoring and Testing*, Lecture Notes in Mechanical Engineering, DOI 10.1007/978-981-10-3247-9_25

contact laser-ultrasonic evaluation (CLUE) system provides pulses with the duration from 10 ns to 100 ns. They can be effectively used for non-destructive testing and material evaluation, namely carbon fiber reinforced composites. Basics of CLUE and some experimental cases are presented in this paper.

2 Basics of CLUE

The schematics of CLUE system is presented in Fig. 1(a). Contact laser ultrasonic evaluation is an ultrasonic testing means employed laser excitation of a smart probe ultrasonic pulse and extra-wide band piezo-detector incorporated into one unit – laser-ultrasonic transducer. Laser pulse, generated by Q-switched diode pumped Nd: YAG laser, is delivered into the laser-ultrasonic transducer through an optical fiber. Probe ultrasonic pulse is excited in a course of laser pulse absorption. It splits into two pulses propagated in opposite direction toward the piezo-detector and a sample, coupled to the transducer with a droplet of liquid. Electric signal of total ultrasonic trace is converted to numeric data, that is processed with PC online.

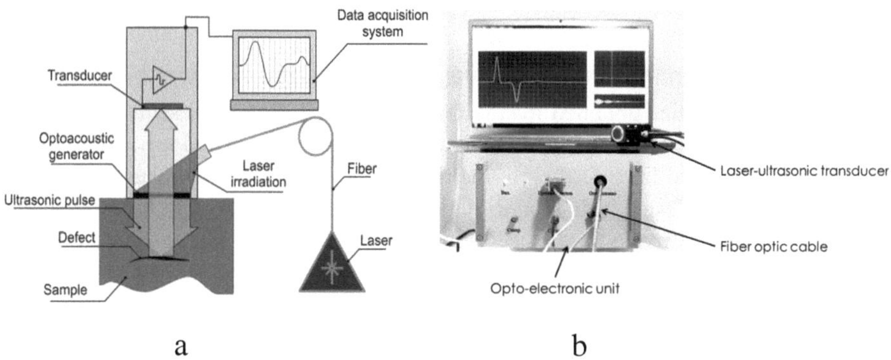

Fig. 1. Schematics (a) and photo (b) of Contact Laser-Ultrasonic evaluation system UDL-2M

The picture of CLUE system for manual operation UDL-2M is presented in Fig. 1 (b). It consists of opto-electronic unit, laser-ultrasonic transducer with the connective cables and PC on-line with the opto-electronic unit. The system can operate with several type of laser-ultrasonic transducer employed for different applications. The laser-ultrasonic transducer for general use PLU-6P-01 is presented in the slide. Its probe pulse has smooth shape without ringing with the width at half maximum of approximately 70 ns, repetition rate is 1 ksps. Short probe pulse duration and its smooth shape provide in-depth resolution of 0.1 mm – in metal and 0.05 mm – in CFRC. Lateral resolution is determined by ultrasonic beam diameter which is $\sim$3 mm. The ultrasonic trace is presented in a screen of PC. The first pulse is a probe pulse, propagated towards the piezo-transducer, the following pulse out of phase with the probe one is the reflection of a probe pulse at the free surface of the transducer. The time delay between them corresponds to twice transit time of ultrasonic wave in the opto-acoustic generator of the transducer. Due to free surface the reflection takes place out of phase, in a case of rigid interface the reflection will take place in phase.

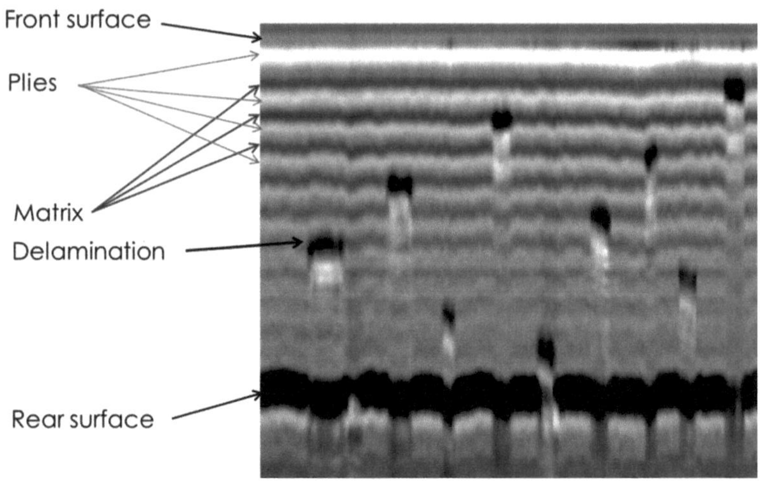

Fig. 2. CLUE of delaminations between the successive layers in the 10-ply carbon-fiber reinforced composite

The B-scan of 10-ply CFRC plate with the delaminations in the successive interfaces between the plies obtained manually with the UDL-2M is presented in Fig. 2.

At the interface matrix-ply the reflection takes place in phase and it is manifested in B-scan as bright solid line. At the ply-matrix interface the reflection takes place out of phase and it is manifested in B-scan as dark or gray solid line. As can be expected the delaminations (black marks) are placed in a position of matrix. The resolution enables evident detection of the delamination between the first and the second ply and successive ones till between 9-th and 10-th. A prominent trace following the delamination at greater depth and corresponding absence of rear surface reflection proves the failure of texture being delamination rather than extra quantity of resin between the plies. The contrast of the signal of texture failure means greater absolute value of coefficient of reflection out of phase relatively to the reflection at regular matrix layer. This is additional reason to treat this failure as delamination.

3 Experimental Cases

The user interface of the CLUE system UDL-2M with the results of CFRC sample manual scanning is presented in Fig. 3. The picture shows both A and B-scans of delaminations in thicker CFRC sample. In addition multiple reflections, taking place in the layer between the front surface and delamination, are evidently manifested. The difference of the amplitude of the pulse reflected at regular matrix layer and failure (see A-scan) proves the last to be delamination. The CLUE data makes it possible to calculate the number of plies in the composite –20 and determine the thickness of the sample –5.37 mm. Numerical filter (the window in right bottom corner) limited frequency band of ultrasonic testing at 0–21 MHz. The exponential decay of the amplitude of the pulse reflected at the successive layers of resin between the plies

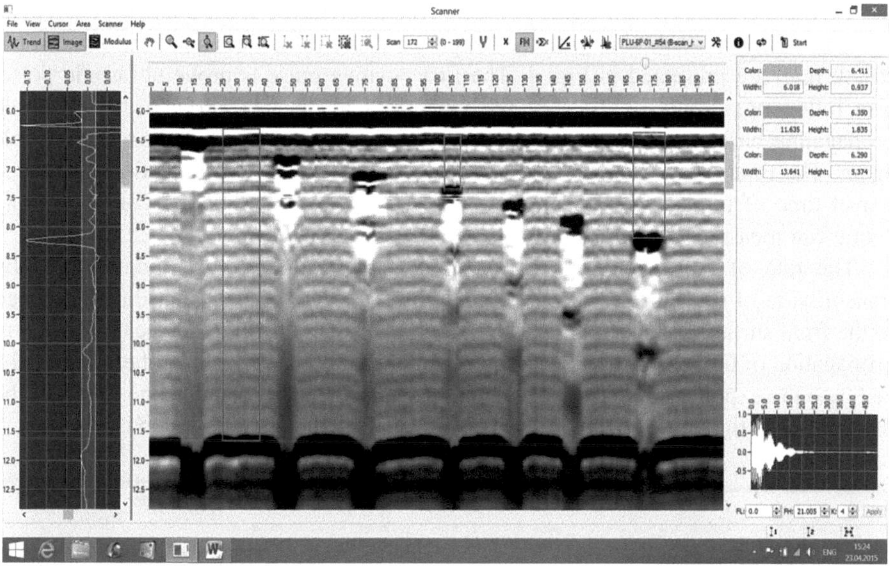

Fig. 3. CLUE of delaminations between the successive layers in the 20-ply carbon-fiber reinforced composite

(see A-scan in the left panel) completely corresponds to theoretical model of ultrasonic pulse propagation in periodic medium and proves high quality of composite texture in regular area.

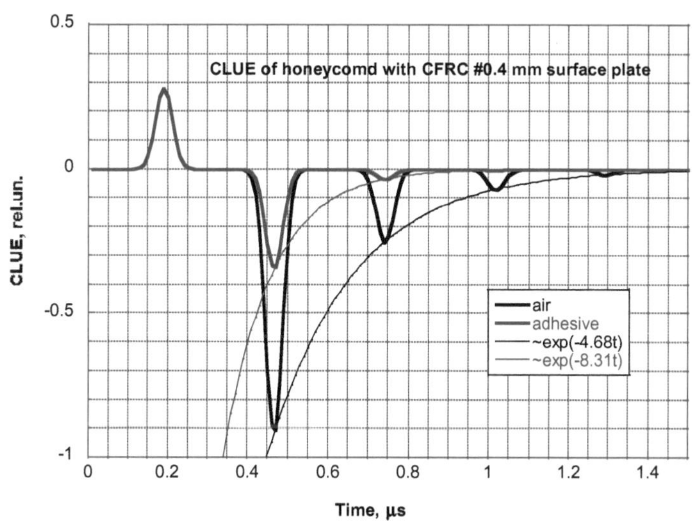

Fig. 4. Decay of successive reverberations in a CFRC cover of honeycomb structure. Blue line – regular adhesion between the covering and inner structure, black line – disbanded covering. (Color figure online)

Theoretical traces of ultrasonic pulse reverberation inside 0.4 mm thick CFRC plate in a case of regular adhesion with the deeper structure (blue line) and lack of adhesion (black line) are presented in Fig. 4. The first positive pulse is probe ultrasonic pulse. The following negative pulses are the reflections at the rear surface of the sample propagating back towards the transducer and reflected back to the bulk of the sample again. The time delay between the appearance of the successive reflections are twice transit time of ultrasonic wave in the sample. Knowing the thickness of the sample L one can measure the ultrasonic wave velocity C_0 in the sample.

The rate of pulse amplitude decay in a case of disbonding is determined by transmission of the successive reflections into the laser-ultrasonic transducer, coupled to the front surface of the sample. The regular adhesion enhances the decay rate due to propagation of the pulses into the structure connected to rear surface of the layer. In the last case the exponent of decay is 1.8 times greater, then for previous one. This can be used for investigation of disbonding in honeycomb structure (see Fig. 5).

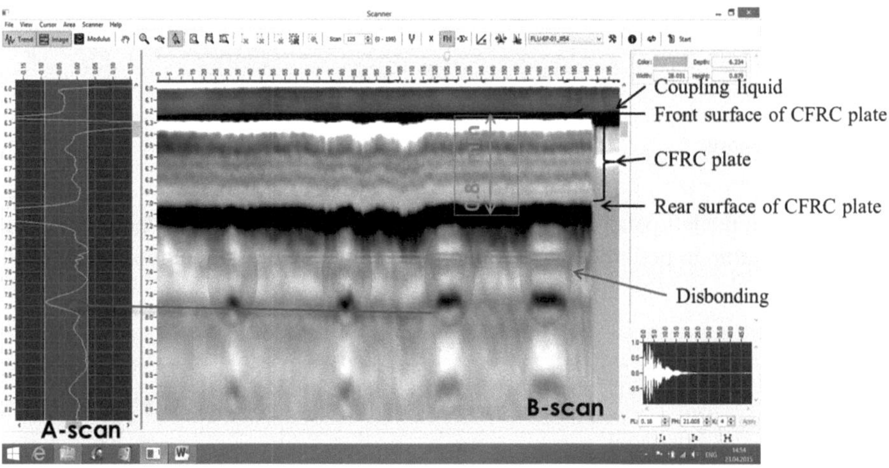

Fig. 5. CLUE of a honeycomb structure with disbanded CFRC cover.

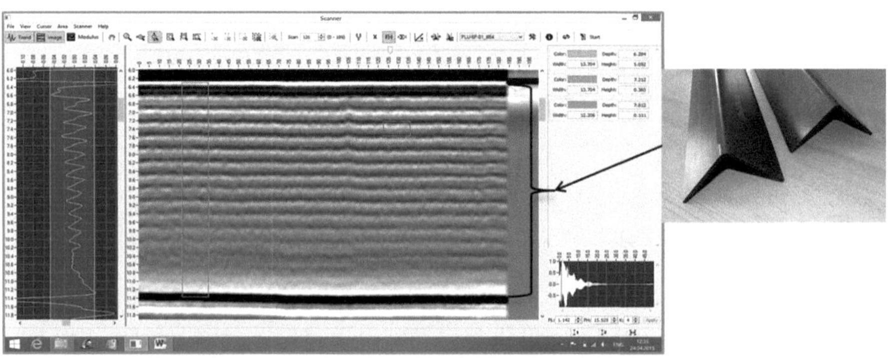

Fig. 6. CLUE of a CFRC part VZ2691544 #028.

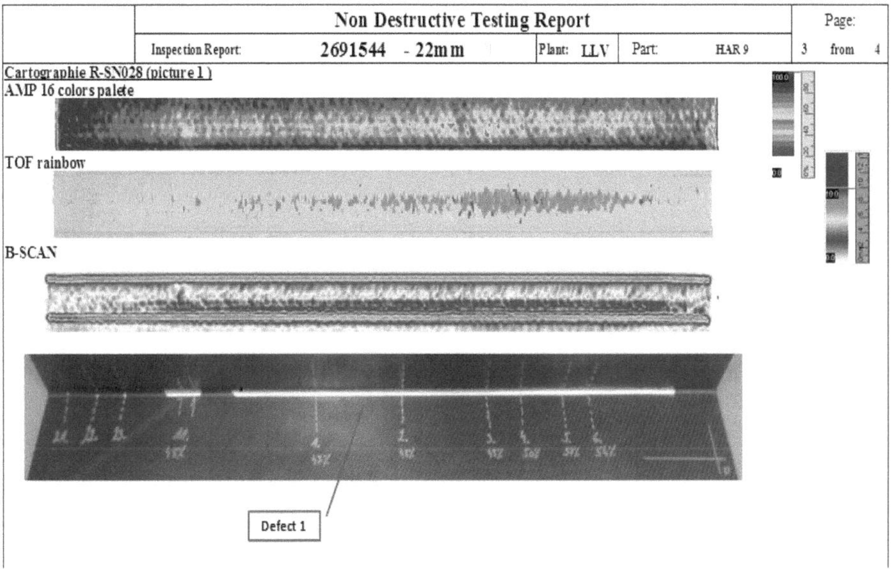

(a)

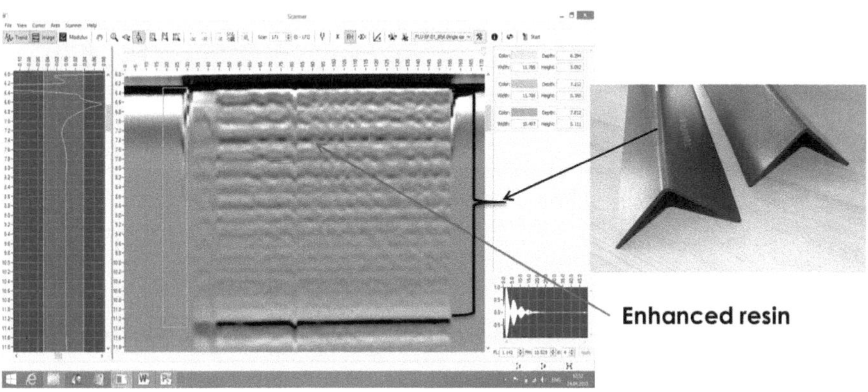

(b)

Fig. 7. Test of the comb of the part VZ2691544 #028 with phase-array in TOF regime (a) and CLUE (b).

B-scan of honeycomb sample with artificial disbonding (see Fig. 5) shows successive reflections in a 0.88 mm thick CFRC cover in the structure failure area. The bright image of the front surface of the sample shows the roughness of the CFRC covering. Black line on the top of the front surface is the image of coupling liquid (water) between the laser-ultrasonic transducer and the sample. High amplitude of the reflected pulses (see A-scan) proves the failure to be disbonding.

The picture of angle CFRC parts tested with CLUE and corresponding B-scan is presented in Fig. 6. The plane part has regular texture without significant irregularities. The thickness of the individual layer of periodic structure of composite is 0.36 mm with the thickness of resin layer $\sim$0.11 mm.

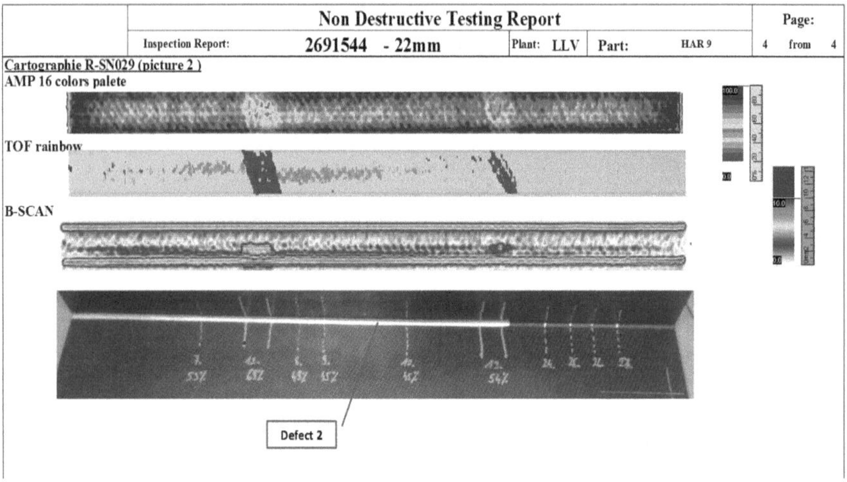

(a)

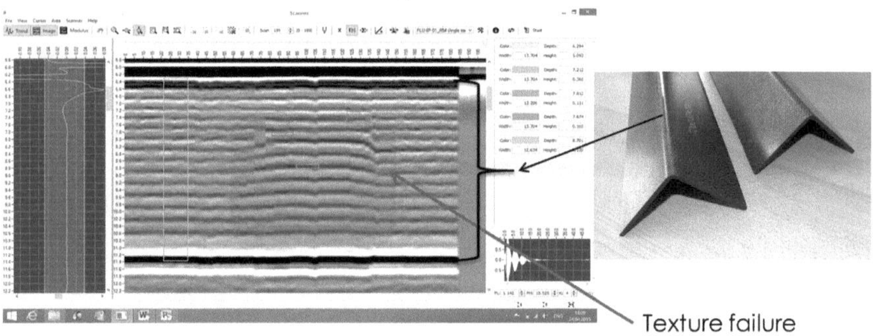

(b)

Fig. 8. Test of the comb of the part VZ2691544 #029 with phase-array in TOF regime (a) and CLUE (b).

The structure of composite in the comb of the part#028 is presented in Fig. 7. The TOF image of the comb is depicted in Fig. 7a. Some irregularities are in the depth from 4 mm to 2 mm from inner surface of the radius. At C-scan is visible back wall echo decrease because of indication and this decrease is to 50 to 60% of FSH. The nature of this failure can be easily specified in B-scan of this part with CLUE (see Fig. 7b). The comb contains extra volume of resin between the third and fourth plies. Some extra resin is presented between the following three plies. This gives rise to the greater reflection of ultrasonic wave at corresponding layers, which is detected by phase array.

The structure of composite in the comb of the part #029 is presented in Fig. 8. The TOF image of the comb is depicted in Fig. 8a. Significant failure of the composite texture is manifested both in TOF and phase-array B-scan (see Fig. 8a). At TOF record is visible significant echo in depth about 3 mm in the same areas where BWE decreased. These indications are visible also at phased array B-scan but nature of this failure can be again specified in B-scan of this part with CLUE (see Fig. 8b). The comb contains flaw of the sixth ply. Some extra resin is presented between the deeper plies. This gives rise to the greater reflection of ultrasonic wave at corresponding layers, which is detected by phase array.

4 Conclusions

The presented results show the enhanced in-depth resolution of CLUE with conventional ultrasonic testing technologies. It provides close to exact information on CFRC texture and properties even in echo-mode with normal incidence of ultrasonic beam. Enhanced in-depth resolution and smooth phase front of laser-excited ultrasonic beam reduces "death-zone" ultrasonic testing to the value of in-depth resolution. It can be as small as 0.05 mm for CFRC with the bandwidth limited to 6 MHz at −6 dB level.

CLUE can be effective tool in a production line and in a structure health monitoring.

Acknowledgements. The work was carried out with financial support in a part from the Ministry of Education and Science of the Russian Federation in the framework of Increase Competitiveness Program of NUST «MISiS» (№ K1-2015-025).

References

1. Scott, I.G., Scala, C.M.: A review of non-destructive testing of composite materials. NDT Int. **15**, 75–86 (1982)
2. Ferreira, A.D.B.L., Nóvoa, P.R.O., Marques, A.T.: Multifunctional material systems: a state-of-the-art review. Compos. Struct. **151**, 3–35 (2016)
3. D'Angelo, A., Bartalini, O., Bellini, V., Levi Sandri, P., Moricciani, D., Nicoletti, L., Zucchiatti, A.: Generation of compton backscattering X-ray beams. Nucl. Instrum. Methods Phys. Res. A **455**, 1–6 (2000)
4. Vavilov, V., Maldague, X., Dufort, B., Robitaille, F., Picard, J.: Thermal nondestructive testing of carbon epoxy composites: detailed analysis and data processing. NDT E Int. **26**, 85–95 (1993)
5. Li, Y., Yang, Z., Zhu, J., Ming, A., Zhang, W., Zhang, J.: Investigation on the damage evolution in the impacted composite material based on active infrared thermography. NDT E Int. **83**, 114–122 (2016)
6. Katunin, A., Dragan, K., Dziendzikowski, M.: Damage identification in aircraft composite structures: a case study using various non-destructive testing techniques. Compos. Struct. **127**, 1–9 (2015)
7. Raišutis, R., Kažys, R., Žukauskas, E., Mažeika, L.: Ultrasonic air-coupled testing of square-shape CFRP composite rods by means of guided waves. NDT E Int. **44**(7), 645–654 (2011)

New Trends Used for Negotiation Untrafficable Terrain

Klára Cibulova(✉) and Jan Sobotka

Department of Engineer Technology, University of Defence,
Kounicova 65, 66210 Brno, Czech Republic
{klara.cibulova, jan.sobotka}@unob.cz

Abstract. This article describes the results of measurement of new perspective material for negotiation low endurable terrain. It is based on the previous measurement with these materials, but enhanced by another vehicles and locations. There are described instruments for measuring the trafficability. The first one is the telescopic penetrometer used by the Czech Army and the second one is the cone penetrometer used by some NATO countries. The main aims of this article are the possibilities of overcoming low endurable terrain with perspective materials – mats. These mats were used during terrain measurements in extremely adverse conditions. It was tested weather the vehicles mired in terrain are able to pass with the mats. The procedures, the calculations and the results of the experiment are presented here.

Keywords: Trafficability of terrain · Telescopic penetrometer · Cone penetrometer · Corps of engineer · Mat

1 Introduction

Trafficability consists of different types of trafficability – the trafficability of woods, waters, soils, towns etc. This means that the trafficability is a very wide and general problem. This article deals with the trafficability of soils. The first aim of the experiment was to determinate the accuracy of an evaluation the trafficability of terrain in highly adverse conditions. The secure of the mobility is one of the tasks of the Corps of Engineers – both the engineer support of troops as well as the assistance for civilians. During the movement the troops do not use only road, but they have to move on the field and on different types of terrains too. These areas are not commonly used for these purposes and that is why the trafficability has to be evaluated. The diagnostic evaluation of terrain is carried out during fulfilling the humanitarian tasks too. So it leads to the possibility and speeding up the rescue works.

The second aim was to suggest the way of negotiation this untrafficable (impassable) terrain. In the Czech Army we have auxiliary types of roads – the permanent and the temporary roads. The permanent roads are roads from soils, geotextiles, branches and wood. The temporary could be from wood and from metal. Due to the demands of time, material and people it begins to think about new and perspective materials – mats. The measurements with these mats are described here.

A. Kravcov et al. (eds.), *Durability of Critical Infrastructure, Monitoring and Testing*,
Lecture Notes in Mechanical Engineering, DOI 10.1007/978-981-10-3247-9_26

The knowledge of trafficability is really so important, because the mobility one of the most substantive assumption of the success of the troops.

2 The Measurement

Conditions of the Measurement. As it was mentioned above this measurement follows the previous one realized with these new materials [1]. This time it was chosen more vehicles and new localization. Although the procedures of measuring the trafficability of terrain are the same, the tests with the mats are new and the results has not been published yet. The measurements took place at the military training area Hradiště. The new localization was chosen – it was a meadow with high humidity, to be able to simulate the extreme conditions.

The tested traces were long 24 m for the Tatras and 8 m for the UAZ.

Instruments used during the experiment were the telescopic penetrometer PT-45. It was chosen because it is used in the Czech army. To be able to get more results it was used cone penetrometer or more precisely it is the trafficability test set comes from USA army and used by some NATO countries. The trafficability was measured in one meter distances by both instruments. The working procedure, differences and the results of the measurements are described here.

Three types of different vehicles were chosen. The first one was Tatra 810 and the second was Tatra 815 and the third was UAZ. The reason was to have trucks and passenger car. These vehicles are a good representation of vehicles used in the Czech army.

The humidity was measured at each one meter on the route. The value of the humidity at the place of mire was from 64–92%. Also soil samples were taken to be classified. They were tested and elaborated at the soil laboratory at the Department of Engineers at the University of Defence. Based on these results the soil was classified as clay sand (SC) Fig. 1.

Evaluation of the Terrain with the Telescopic Penetrometer. As it was mentioned above this is Czech instrument (used by Czech engineers). The procedure of working is dealt with the Field Manual "Military Roads and Ways" [2]. Due to this manual, it is possible to judge the trafficability by several methods. It could be mentioned by visual observation, making a footprint on the soil, an engineer crowbar and a telescopic penetrometer (PT–45). The two latest are based on finding out the resistance of soil against leaking of the thorn. The most accurate is the telescopic penetrometer and that is why it was used in these measurements Fig. 2.

The telescopic penetrometer is ended with a thorn; this thorn is pressed into the soil. The instrument has a dial on which we read the pressure needed to press the thorn to different depths. Each measuring is carried out three times in one-meter distance. The number of vehicles, which can negotiate the measured area, is determined due the table in the Field Manual. As a low-supporting terrain we consider the terrain where the telescopic penetrometer measures the value lower than 3 MPa [2] (Fig. 3).

Evaluation of the Terrain with the Cone Penetrometer. The cone penetrometer comes from the Army of United States and it is used by some other NATO countries as well. That is why it was chosen to the experiment. The procedure of working is dealt

Fig. 1. Soil sampling.

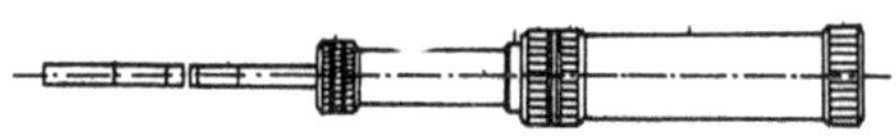

Fig. 2. The telescopic penetrometer PT-45 [2].

with the Field Manual "Field Manual 5-430-00-1 Planning and design of roads, airfields, and heliports in the theater of operations – road design" [3]. In this field manual we determine the trafficability of the soils by two indexes – rating cone index (RCI) and vehicle cone index (VCI). As soon as we know the values of these two indexes and compare them, we are able to judge if the soil is trafficable for the given number of vehicles.

Rating cone index - RCI is measured with the soil-trafficability test set. RCI is a product of two other indexes - cone index CI and remoulding index RI [3].

The CI is measured with the cone penetrometer. The amount of force required to move the cone slowly through a given plane is indicated on the dial inside the ring. This force is an index of the soil's shearing resistance and is called the soil's CI in that plane. The index RI is measured with this set as well. A piston-type soil sampler is used to extract soil samples for remolding tests (Fig. 4).

For conventional types of vehicles used in some NATO countries, the values of vehicle cone indexes are known and tabulated. On the other hand, we have to count the

Fig. 3. Measuring with the telescopic penetrometer.

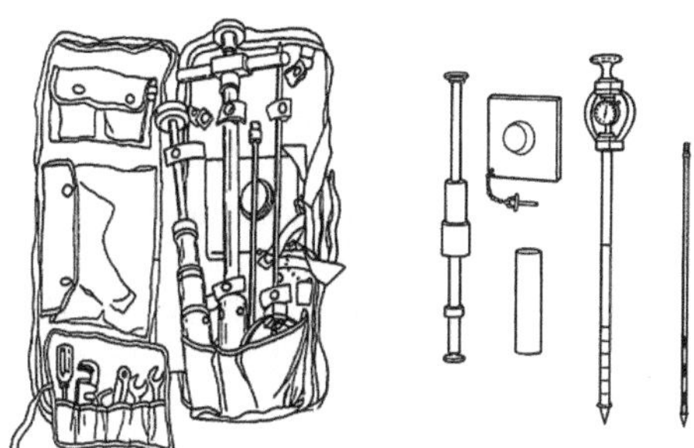

Fig. 4. The trafficability test set [3].

VCI for the vehicles used in the Army of the Czech Republic. First we are able to count the mobility index MI and after that we can find the vehicle cone index from a curve in a chart for one or fifty passes. Here is the formula for counting MI [3]:

$$MI = \left[\frac{(CPF) \times (WF)}{(TF) \times (GF)} + (BF) - (CF)\right] \times (EF) \times (TSF), \tag{1}$$

where: *PF* - contact pressure factor, *WF* - weight factor, *TF* - track factor, *GF* - grouser factor, *BF* - bogie factor, *CF* - clearance factor, *EF* - engine factor, *TSF* - transmission factor [3] (Fig. 5).

Fig. 5. Measuring with the cone penetrometer.

Results of Evaluation of Terrain. The route was firstly measured with both penetrometers and it was evaluated, whether the vehicles are able to negotiate it or not. Presupposes due to the results of the penetrometer measurements are show in Table 1. There are shown the distances of real mires of both vehicles as well. These really extremely adverse conditions caused that the vehicles mire (Fig. 6).

It can be seen that the whole route was 24 m for Tatras and 8 m for UAZ. The results are following.

First measurement was made by telescopic penetrometer PT-45. Due to the evaluation of this instrument the route is trafficable for the Tatras to the distance 3 m and for the UAZ to the distance 4 m.

The second measurement was made by the cone penetrometer – the route was trafficable for the Tatras to the distance 6 m and for the UAZ to the 4 m.

Then the humidity was measured – as well as the penetrations each measurement was made at one meter distances. The soil sampling followed.

The passing of vehicles followed. The first went the Tatra 815 and it mired at the distance 16 m, Tatra 810 mired at 12 m and UAZ at 6 m.

One thing, which was not been taken into consideration is the skills of the driver. These consequences we are not able to measure, but unfortunately they are not negligible.

But finally it has to be said that the measurements always corresponded to the real situations and the trafficability was evaluated correctly on the safety side.

Table 1. The results of measuring with the penetrometers.

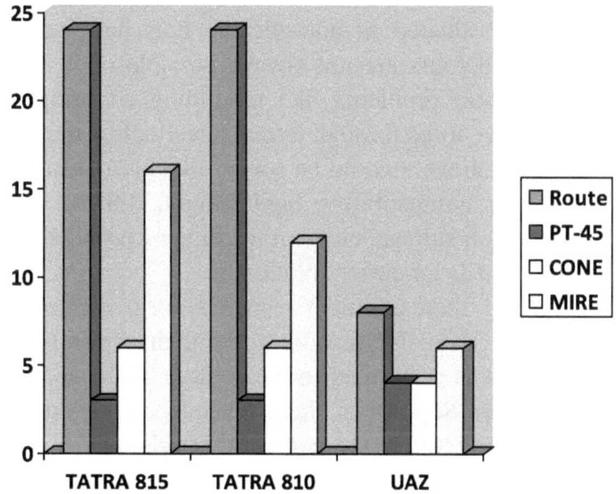

Fig. 6. The mire of vehicle Tatra T-815.

3 New Trends Used for Negotiation Untrafficable Terrain

In the case the route is evaluated as untrafficable it is necessary to solve how to negotiate the terrain. The detours are not always possible or it is too long. In critic situations there are common problems like rebuilding of bridge or reparation of damaged object, so you have to go through terrain nevertheless there is not a functional road. As it was mentioned above it could be solved like permanent or temporary road. The problem is with their compatibility, high weight, difficult transportability and maneuverability, demands on storage, etc., but the biggest problems are material, labor and time demands. So what is an easier solution?

Currently on the market, there are many resources for overcoming the untrafficable terrain. The analysis of the available resources and their use in the conditions of the Army of Czech Republic was performed in the context of the specific research [4] of the Department of Engineer Support of the University of Defence. On the basis of multi-criteria evaluation, the tools of the company Deschamps were selected as being the most appropriate. These are special plastic mats (based on polyester fibers) with high load capacity and resistance designed for increasing the load capacity of the terrain (the construction of auxiliary routes and roads, heliports, etc.) These mats are already implemented and used in some NATO armies. They are not yet implemented and even have not been officially tested in the Army of the Czech Republic and for such purposes like negotiation low endurable terrain - especially clay and mould.

These Deschamps mats were used during measurements of trafficability of terrain in the military area of Hradiště. The tools of the Deschamps Company are supplied in different variants [5]. These variants were available for testing: Traction-Matt (two rolls of a width of 0.8 m and a length of 8.0 m) and Mobi-Mat MM3V25 (a width of 2 m and a length of 5 m). The selected location was based on penetrometrical measurements evaluated as unable to negotiate, which were subsequently proven by stuck vehicles (see Table 1). Plastic mats were subsequently put in places where the vehicles got stuck Fig. 7. The vehicles were able to continue in overcoming obstacles, by at least one axle, after accessing the mats. Unfortunately, it was not possible to cover the entire obstacle with mats so that the evaluation could be completely conclusive. Therefore, it would be appropriate to continue with the testing, but to ensure the greater dimensions of the plastic mats and various types of load capacities and also to expand the types of vehicles tested.

The Table 2 shows the whole route –24 m for Tatras and 8 m for UAZ. It can be seen the places of mire – Tatra 815 mires at distance 16 m, the Tatra 810 at distance 12 m and the UAZ at distance 6 m. The most important which is shown in this table is the possibilities of overcoming this untrafficable terrain. After the vehicles mire the mats were laid on the place of miring and as you can see then the vehicles were able to negotiate this untrafficable terrain with this mats.

Fig. 7. The negotiation of terrain with the mats.

Table 2. The results of negotiation of terrain with mats.

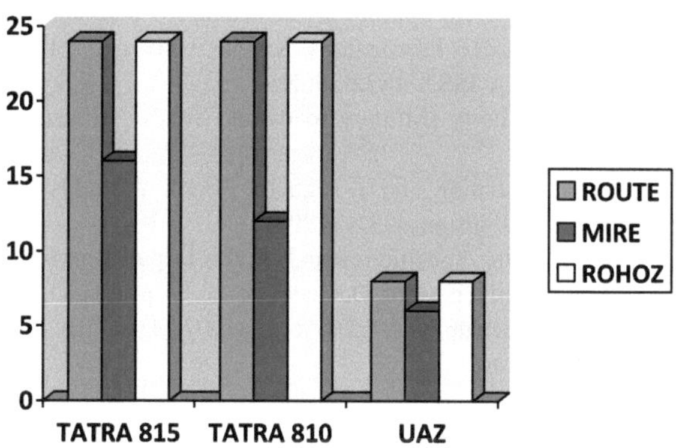

4 Summary

Military engineering of mobility is a key factor in the framework of the task fulfilment of the Corps of Engineers on the territory of the Czech Republic as well as on foreign missions. Fast and accurate determination of the terrain trafficability is an essential prerequisite for the subsequent solution of mobility of units. The chosen area represented the extreme conditions of possible crisis situations very well. The measurements taken by penetrometers (a PT-45 telescopic penetrometer and cone penetrometer) clearly classified the terrain as untrafficable. The relevance of the measurement was confirmed by a subsequent experiment, in which the selected vehicles got stuck – mire.

In the experiment, it was verified that the terrain evaluated as untrafficable is possible to transform to trafficable with the use of plastic mats (simple and non-time-demanding tools). Currently, Czech Corps of Engineers use wooden or steel structural elements of various parameters for increasing the load capacity of the terrain requiring a large amount of material, equipment and labor. The replacement of these tools for the modern and commonly used ones in the armies of NATO will contribute to speeding up and simplifying the tasks provided by engineers, but also other troops of the Armed forces of the Czech Republic.

But still there were performed only few measurements with only chosen types of mats. The recommendations are to perform more measurement with more types of mats and vehicles and at different locations to be able to evaluate more dates.

Acknowledgment. The work presented in this article has been supported by the Czech Republic Ministry of Defence - K 201 Department development program "Development of technologies in weapons construction, ammunition, weapons instrumentation, materials engineering and military infrastructure".

References

1. Cibulová, K., Sobotka, J.: Different approaches of evaluation and negotiation low endurable terrain. In: Transport Means 2016 Proceedings, Kaunas University of Technology, Kaunas, Lithuania, pp. 241–244 (2016). ISSN 1822-296X
2. Žen 2-16 Vojenské silnice a cesty [Military roads and Ways]. Ministry of Defence, Praha, 226 p. (1987)
3. Field Manual 5-430-00-1 Planning and Design of Roads, Airfields and Heliports in the Theater of Operations – Road design. USA (1994)
4. Kyjovský, J.: Final report of the Specific research of the Department of Engineer Support of the University of Defence. University of Defence, Brno, 34 p. (2011)
5. Dechamps. La Couronne: Deachamps (2015). Accessed 07 May 2016. http://www.mobi-mat-aircraft-recovery-deschamps.com

The Protection of Critical Infrastructure Objects – Technical Principles

Pavel Maňas(✉)

University of Defence,
Kounicova 65, 662 10 Brno, Czech Republic
pavel.manas@unob.cz

Abstract. The protection of critical infrastructure is not only the question of their resilience but in a technical point of view, it is a question of protection of function of particular objects of critical infrastructure. This protection of objects depends on possible threats which are heavily to determine, one of possible way is to use commonly agreed procedures like e.g. STANAG 2280. Possibilities in assessment of structures facing to attack are described and examples of simulation are given.

Keywords: Critical infrastructure · Protection · Threads · Blast · Vehicle impact · FE simulation

1 Introduction

Basic principles and core ideas about critical infrastructure protection is addressed in several official papers of EU or in national guides or directives [1]. At this level, critical infrastructure is solved sector by sector or in interaction between sectors. There exists common sign between sectors, critical infrastructure includes inter alia objects with characteristics similar to any other structures:

- design and shape of objects are determined by function,
- objects have structural stiffness, robustness and integrity.

As the core of critical infrastructure protection is keeping its functionality, the robustness and structural integrity is in the focus. Each key object of critical infrastructure has to be properly designed with robustness and reliable resistance against main effects of attack or proper measures focusing on resilience and robustness have to be done. The type of attack or threat is discussed later on this paper.

Depending of a scale and actors, the critical infrastructure elements should be considered e.g. in European, national, NATO or national military context, where we can speak about "critical" infrastructure elements for deployed forces during an operation. In the military context, the "critical" infrastructure elements are mainly temporary and completely different comparing to permanent national or regional critical infrastructure. The solutions for these kind of infrastructure elements could be different, but key principles are the same. This leads to idea of protected structures,

A. Kravcov et al. (eds.), *Durability of Critical Infrastructure, Monitoring and Testing*, Lecture Notes in Mechanical Engineering, DOI 10.1007/978-981-10-3247-9_27

which are protected by protective constructions or measures, based on practical experience with blast effects on structures and personnel, or on modern methods such as simulations.

2 Threads and Assessment

The one of the most actual security issue for common population around Europe is a terrorist attack on place with high concentration of people. It is only a question of time, when the critical infrastructure elements will be considered as an attractive target for such kind of attack. In this case, critical infrastructure elements are mainly represented by important buildings or similar objects and thus we are speaking about attack against the function of this building, their structural integrity and occupants of this building doing their businesses.

From technical point of view, it is necessary to answer several crucial questions for each protected structure or protective construction:

- How to define threats;
- How to calculate the action on protective construction or protected structure;
- How to calculate the response of protective construction or protected structure;
- Criteria for assessment.

First bullet is a question for security analyst, intelligence services and other experts, but there does not exist simple answer or mathematical rule how to determine mass and type of charge, which could be used for following calculations, every time there is

	A Small / medium calibre projectiles	B Shoulder launched weapons / Rifle grenades	C Battlefield rockets, Artillery and Mortars	D Small / Personnel-borne IEDs	E VBIEDs
5	Automatic cannon 30 mm APDS	Advanced ASM Anti Structure Munition	155 mm artillery 122 mm rocket	Bag / Suitcase 20 kg TNT	Heavy truck / similar > 4000 kg TNT
4	Heavy machine gun 12.7 – 14.5 mm AP	Anti-tank Shaped charge	120 mm mortar 107 mm rocket	Body-borne device 9 kg TNT, fragments	Medium truck 4000 kg TNT
3	Assault / Sniper rifle 7.62 mm AP WC	Anti-personnel Thermobaric charge < 2.5 kg / Conventional	82 mm mortar	Large briefcase 9 kg TNT	Van 1500 kg TNT
2	Assault rifle 5.56 – 7.62 mm AP	40 mm Rifle grenade Shaped charge	60 mm mortar	Package 1.5 kg TNT	Passenger vehicle 400 kg TNT
1	Assault rifle 5.56 – 7.62 mm Ball	(Reserved)	Hand grenade	Letter bomb 0.125 kg TNT	Motorbike 50 kg TNT

Fig. 1. Design threat levels [STANAG 2280:2008 Ed. 1]

narrow or wider interval of values, which should be estimated. Therefore, in a protective measure against effects of blast attack does not exist one correct solution, but complex measures covering above mentioned interval. One of possible way how to address this issue is using of widely accepted matrix of design threats levels (see Fig. 1), which is agreed and used among militaries in NATO (Table 1).

Next two bullets can be addressed by using of physical laws, solution is based either on essential formulae taking into account mass of charge and distance from epicentre [3] suitable for hand calculation or on numerical simulation using special solvers based on explicit solution of motion equations, which need really powerful computers [2, 4, 5].

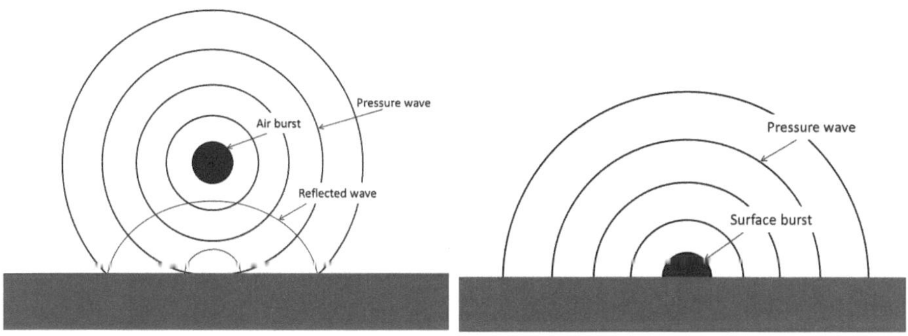

Fig. 2. a: Spherical free air burst. b: Hemispherical surface burst

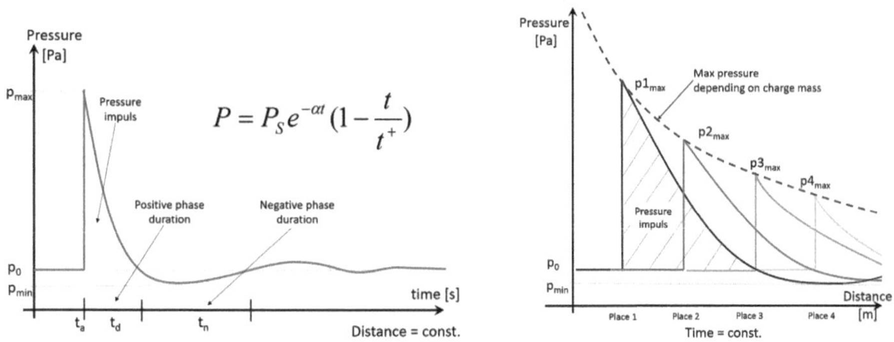

Fig. 3. a: Dependency of overpressure at given distance on time of duration after explosion. b: Chart of maximum pressure at distance from explosion

To calculate basic data for assessing a structure, three main methods can be used:

- *Empirical calculation* based on scaled distance from TNT charge and empirical formulae of overpressure, pressure impulse and time of arrival and time of duration [3];

Table 1. Damage criteria for structures or components due to overpressure – examples [kPa]

Damage/Object	Minor	Medium	Heavy	Destruction
	Overpressure Δp [kPa]			
Concrete wall, 20–30 cm	–	–	14–21	–
Brick wall – completely demolished	–	–	56.3	70.3
Brick wall, 20–30 cm – fail by flexure	–	–	–	56.3
Brick wall, 45 cm – completely demolished	–	–	–	91.4
Steel building	9.1	14.0	17.6	21.1
Wooden building	–	12.0	17.0	28.0

- *Semi-empirical calculation* based on the same formulae like previous but for simple geometries, some software tools can be used for calculation (e.g. Con-Wep and BlastX from U.S. Army Corps of Engineers, Protective Design Centre, or Mathcad handbook DynamAssist);
- *Numerical method based on explicit solution* of motion equations, appropriate solver is often part of complex simulation software and can solve complex geometry and loading conditions.

Empirical calculations of blast effect are mainly focused on spreading of pressure wave in the air and calculation of overpressure (see Fig. 3) for both types of burst – air burst and surface burst (see Fig. 2).

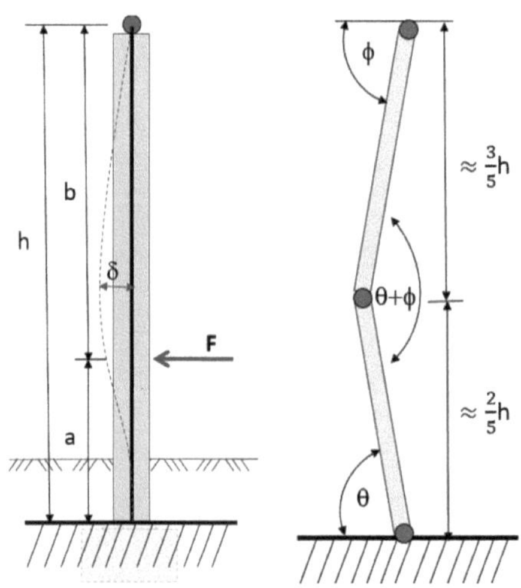

Fig. 4. a: Bridge column after impact of truck. b: Schema of impact of vehicle into bridge column [7].

Fig. 5. a: Damaged protective wall after VBIED attack. b: Simulation of blast behind T-Walls, 25 ms after detonation of TNT charge

From practical point of view, we can consider each bomb attack through way of delivery of weapon to spot (see Fig. 1, column D, E) as a hemispherical surface burst (see Fig. 2b), where the maximum overpressure in the front of blast wave is about two times higher than the same but air burst. Following table brings some examples of overpressure which can cause damages on typical structures [4] (Table 1).

The assessment criteria are problematic again, because there does not exist simple and reliable criteria similar to Eurocodes for designing of structures, which could be easily used for assessment of damages on particular structures. There are a lot of papers [e.g. 4–6], where are tables with overpressures of blast waves and its possible damages on some kind of structures, but this is suitable for managers, not for structural engineers. Another important point is, that not overpressure but pressure impulse is primer source of damages on structures.

Different type of possible attack can be an intentional impact of vehicle into the protected structure or into perimeter barrier aiming to broke into denial area and later explode or disrupt the function of protected infrastructure directly (see Fig. 6).

Solution of that problem can be obtain with use of basic physical laws [7] or numerically, where the correct stiffness behavior of column – rigid obstacle and vehicle – moving mass is taken into account (Fig. 4).

3 The Role of Simulation

The analytical solution is suitable for quick guess or simple geometry of target and charge or vehicle, the advantage is weak dependency on material characteristics, the disadvantage is low reliability or preciseness of obtained results. On the other hand, numerical solution can give us almost correct output data, but strongly dependent on quality of material model, input values and complexity of whole numerical model of the event and it is really time consuming.

6a Blast of 1000 kg of ANFO – overpressure in the front of pressure wave [kPa] - time 0 ms

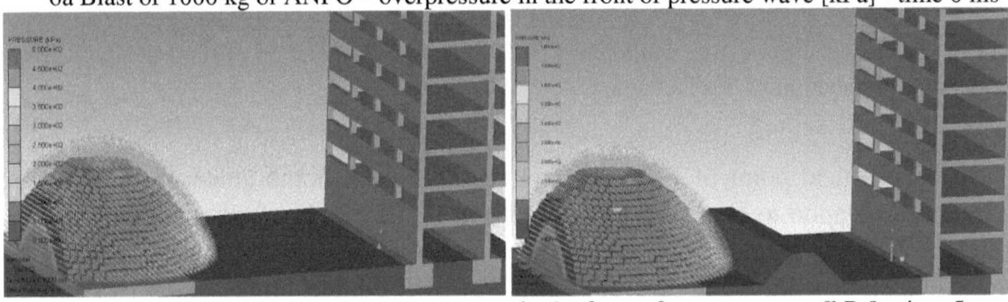

6b Blast of 1000 kg of ANFO – overpressure in the front of pressure wave [kPa] - time 5 ms

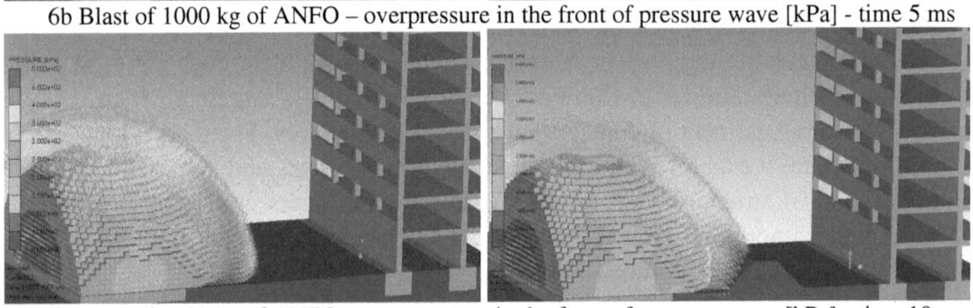

6c Blast of 1000 kg of ANFO – overpressure in the front of pressure wave [kPa] - time 10 ms

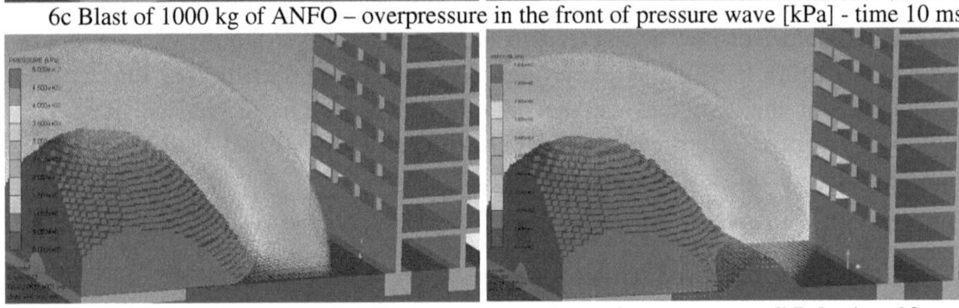

6d Blast of 1000 kg of ANFO – overpressure in the front of pressure wave [kPa] - time 20 ms

Fig. 6. a Blast of 1000 kg of ANFO – overpressure in the front of pressure wave [kPa] - time 0 ms. b Blast of 1000 kg of ANFO – overpressure in the front of pressure wave [kPa] - time 5 ms. c Blast of 1000 kg of ANFO – overpressure in the front of pressure wave [kPa] - time 10 ms. d Blast of 1000 kg of ANFO – overpressure in the front of pressure wave [kPa] - time 20 ms. e Blast of 1000 kg of ANFO – overpressure in the front of pressure wave [kPa] - time 30 ms. f Blast of 1000 kg of ANFO – overpressure in the front of pressure wave [kPa] - time 40 ms. g Blast of 1000 kg of ANFO – overpressure in the front of pressure wave [kPa] - time 50 ms

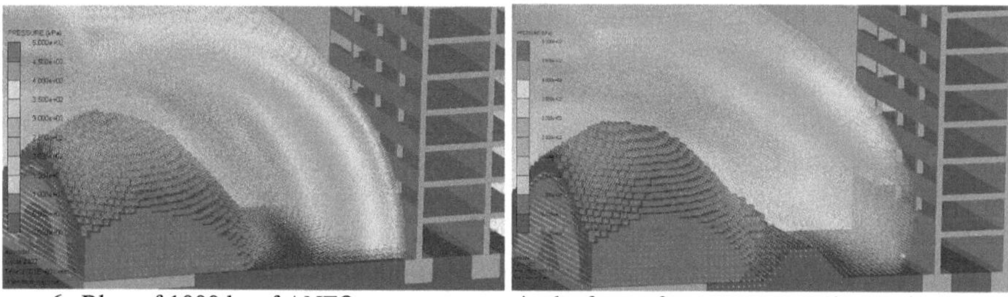

6e Blast of 1000 kg of ANFO – overpressure in the front of pressure wave [kPa] - time 30 ms

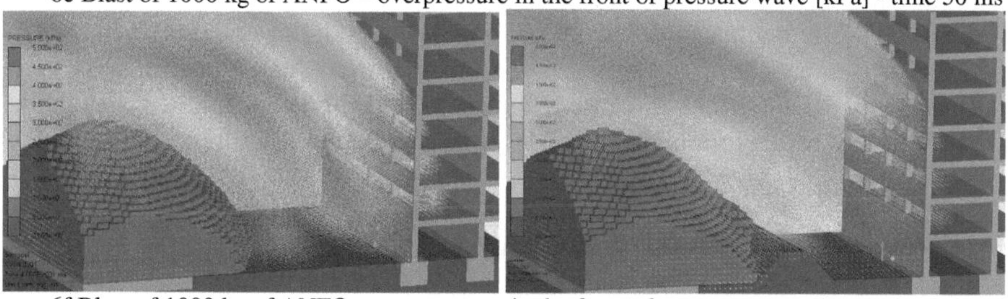

6f Blast of 1000 kg of ANFO – overpressure in the front of pressure wave [kPa] - time 40 ms

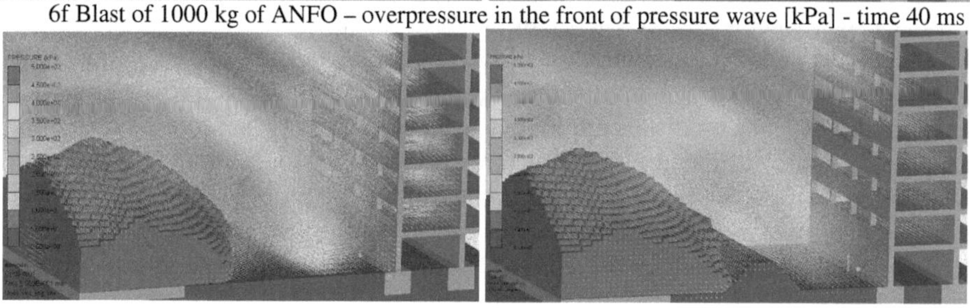

6g Blast of 1000 kg of ANFO – overpressure in the front of pressure wave [kPa] - time 50 ms

Fig. 6. (continued)

Numerical simulation is mainly based on modern finite element method, where is possible to solve numerically the motion equation (Eq. 1) with complex geometry and physics environment.

$$m \cdot \ddot{u} + c \cdot \dot{u} + k \cdot u = F(t) \tag{1}$$

The crucial problem of each numerical simulation is a number of suitable evaluation criteria. It is possible to use numerical simulation of damages due to air blast or impact on structural first members but concerning whole structure the simulation is limited by computer and software limitations - 3D simulation of steel or concrete structural member hit by pressure wave costs millions equations and several days or weeks for solution of 100–500 ms of effects.

Numerical methods based mainly on Euler or Lagrange solver allow us to compute complex simulation where pressure, velocities and deformations are basic output data at each point of simulation and damage levels, strains and other structural characteristics are available too. Euler solver is more suitable to simulate blast wave spreading and interaction with rigid structures. Lagrange solver is effective in simulation of flexural bodies interaction, e.g. impact of vehicle into obstacle or bullet with target. If there is a need simulate interaction of blast wave with real flexural structure, the combination of Euler and Lagrange solver can be used, where blast wave is an Euler domain and flexural structures are Lagrange bodies. Following example (see Fig. 5a, b) compares damage after VBIED attack and simulation, where the combination of Euler and Lagrange solvers were used [4, 5].

The typical calculation time of these type of simulation depends on complexity of problem but range from several hours to several weeks. A significant limitation for all of these tools is in defining realistic failure criteria for both the structural elements and the humans. Depending on the scenario, the failure criteria for a human may be set as the blast pressure threshold to cause a burst ear drum, internal injuries, for another scenario it may be set to a higher blast level to cause significant injury or fatality.

4 Principles of Protective Measures

Blast is a significant release of energy from MJ to GJ during short time (milliseconds), this energy depends on type and mass of charge and cause blast pressure wave in air, which acts on structures (overpressure and pressure impulse). Overpressure and pressure impulse depends on distance from epicenter and shape of pressure wave and they are primary source of damages on structures or occupants of the building. Impact of vehicle into obstacle is similar physical event, kinetic energy in MJ of moving vehicle is transferred into obstacle as deformation energy during less than 0.5 s, this energy depends on mass and velocity of vehicle. Transformation of this energy into physical work is the primary source of damages on structures.

Following Fig. 6a to g shows the principle how to reduce overpressure in the front of blast pressure wave and how to reduce pressure impulse on façade of building. Left side has no obstacle between charge and object, right side has a soil obstacle between charge and building, the charge is the same – 1000 kg of ANFO, stand-off distance is 25 m, first 50 ms is simulated.

The shape of blast wave and overpressure is the same till the wave reach soil obstacle in time 7 ms, then pressure wave start to bend around obstacle – time 10 ms. When the pressure wave come over the obstacle – time 20 ms, the maximum overpressure is almost the same in both cases, but just behind the obstacle, there is significantly lower overpressure at the ground level. When the pressure wave hits the object –30 ms, there is significant difference between pressure level of direct wave on left side and bended pressure wave on right side, in addition, the maximum is not on the ground level but between 2nd and 3rd level. When the pressure wave broke into object through windows – time 40–50 ms, the pressure level is on the right side much more favorable, in addition, the direct wave on left side produce significant reflected pressure impulse causing additional damages on object.

5 Summary

Protection of structures against effects of blast is a complex and expensive issue. Smart design of new buildings or objects, taking into account safety distance from possible epicenter of blast, can save a lot of money and efforts in future adjustment of these objects for new threats. New advanced materials, innovative combination of materials or functional structural elements can help experts design appropriate measures to improve robustness, structural stiffness and integrity of important buildings and objects.

Protecting critical and military infrastructures from explosive and impact attacks:

- Based on physical phenomena – the best solution is a distance resp. lowering the velocity of vehicle
- Resistant structure should be able to dissipate a lot of energy of blast wave or kinetic energy of impacting vehicle (optimized stiffness and materials)
- Hand calculation based on analytical and empirical formulae are integral part for conceptual design of safety measures
- Numerical simulation (FEM) using explicit formulation of motion equation is useful for detail case studies and analyses
- Assessment criteria for damages on structures and humans can be a gap (preciseness, consistent, widely adopted).

Nowadays, materials used for ballistic or blast protection are mostly materials on the metallic base, various non-cementitious composite materials and different kinds of concrete, but there is no universal material suitable as a ballistic protection and protection against blast.

Modern software tools like ANSYS/AUTODYN or LS-DYNA can help experts to properly assess threats in asymmetric conflicts for reasonable cost. Due to the complex nature of the high velocity interaction between bodies or blast wave spreading and physical phenomena being analyzed, it is extremely important for the user of mentioned tools to have a good understanding of the underlying assumptions and limitations of the models.

Acknowledgment. The work presented in this article has been supported by the Czech Republic Ministry of Defence - K 201 Department development program "Development of technologies in weapons construction, ammunition, weapons instrumentation, materials engineering and military infrastructure".

References

1. Information on http://eur-lex.europa.eu
2. Štoller, J., Maňas, P., Zezulová, E.: Blast Testing and Simulation Methods, pp. 35–100. Česká Technika-nakladatelství ČVUT - Praha, Praha (2015). ISBN 978-80-01-05898-5
3. IATG 01.80: 2015. Formulae for Ammunition Management, 2 edn. UNODA (2015)
4. Maňas, P., Kroupa, L., Urban, R., Coufal, D.: Blast threat to critical and military infrastructure. Secur. Defence Q. **2013**(1), 32–53 (2013). ISSN 2300-8741

6. Larcher, M.: Simulation of the effects of an air blast wave. JRC Technical Notes, PUBSY JRC41337 (2007). <ipsc.jrc.ec.europa.eu>
7. Vintr, Z., Maňas, P., Vala, M.: Evaluation of the effects of high speed truck crash into a bridge column. In: Proceedings of the 6th International Conference on Mechanics and Materials in Design (M2D2015), pp. 1025–1036. Universidade de Porto, Porto, Portugalsko (2015). ISBN 978-989-98832-3-9

Using of Intelligent Transport Systems to Elimination of the Negative Effect on the Transport Security

Michal Ballay(✉), Lucia Figuli, and Zuzana Zvaková

Faculty of Security Engineering, University of Zilina,
1. Mája 32, 010 26 Žilina, Slovakia
{michal.ballay, lucia.figuli,
zuzana.zvakova}@fbi.uniza.sk

Abstract. Paper is focused on the important aspects of the deployment and the use of intelligent transport systems in road transport. Intelligent transport system are one of the tool how to increase the security of one of the sectors of critical infrastructure – transport. The paper defines the concept of traffic safety and assesses the current condition of road safety in the European Union. The paper concludes with the design of varying Electronic Bulletin Board, as a tool of intelligent transport systems. It has bring the elimination of negative impacts of transport systems in road transport.

Keywords: Intelligent traffic systems · Security · Road traffic · Variable electronic information board · Critical infrastructure

1 Introduction

The development of transport in all its sectors is linked to the development of society. One of the sectors of critical infrastructure is the sector of transport. One of the subsectors of this sector is road transport. Using of information technologies in the society is also reflected in the management of transport. Control and information systems in transport is becoming essential component of modes of transport in this time. According to the international studies, the implementation of intelligent transport systems (ITS) in traffic can greatly reduce the likelihood of crisis events and can increase the safety of critical infrastructure elements. These sophisticated multimodal interdisciplinary tools using specific tools for each type of traffic, are also the backbone of complex solutions. They have a positive impact on the safety and on the traffic fluidity. Their use can also be applied in the area of crisis management.

2 Term "Safety" in Road Transport

The official World Health Organisation (WHO) safety definition is 'a freedom from unacceptable risk of harm'. Road safety is, however, usually defined in a negative way. Safe road traffic is characterized by the absence of crashes, injuries and fatalities. Crashes are used here instead of accidents because the word accident leads thoughts in the direction of an event that is hard to do something about ¾ "it happened by accident" [1].

A. Kravcov et al. (eds.), *Durability of Critical Infrastructure, Monitoring and Testing*,
Lecture Notes in Mechanical Engineering, DOI 10.1007/978-981-10-3247-9_28

For society and for the individual, the loss of health is the most serious effect of crashes. It leads both to serious personal suffering and to huge societal losses in monetary terms. Therefore, it is essential to state at this early stage that loss of health is the main road safety criterion. Convalescence and health restoration have economic consequences as have vehicle damage and loss of vehicles and goods [1].

Road safety may be influenced along many separate dimensions and various models have been used in road safety management:

- There are three main variables that decide the road safety level in terms of health consequences: exposure in traffic, risk of a crash given the exposure, consequence of the crash.
- The health consequences of road crashes may be influenced by actions taken before the crash (active safety), during the crash (passive safety), and post crash (rescue, treatment, rehabilitation). ·
- Most of the crashes are triggered by human errors, sometimes intentional but normally unintentional. There are three basic ways to reduce the human errors: selection of road users (for example, licensing), improving road users (for example, information, education, training, and enforcement), adaptation of road and vehicle engineering solutions to human characteristics and limitations (make it easier to drive, to bicycle, to walk in traffic) [1, 2].

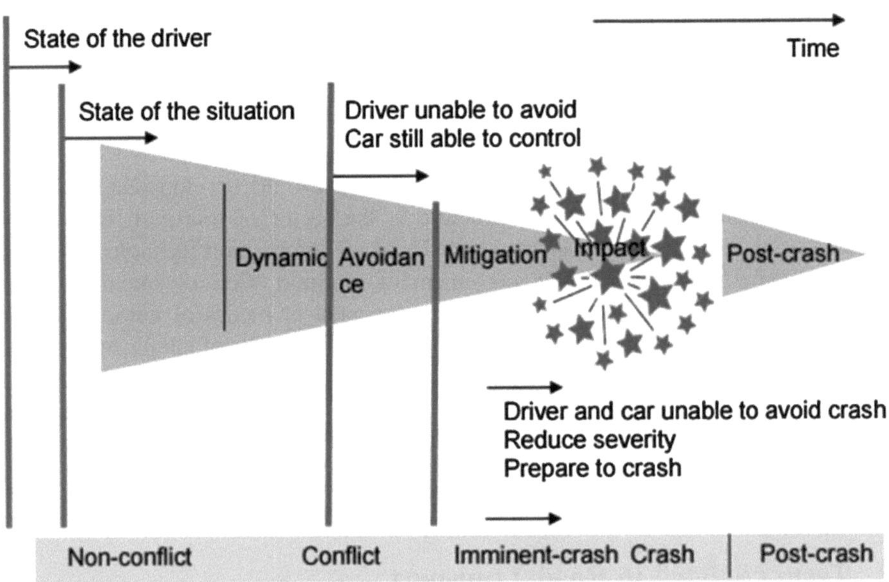

Fig. 1. The holistic view of safety [1]

Transport safety can be defined in a number of ways. However, there are several "specifiers" of this concept. Often is used the terms "**traffic safety**", which, however, is a general term and can refer to the safety of all types of traffic: air, rail, road [3].

Similarly, as another general definition: "safety of transport is the most important principle of each operating transport system; it is a basic component of all metrics of success" [4].

And here is another interpretation of "traffic safety": "*Term that is related to the negative performance of the traffic system to generate traffic accidents that involve injury or fatality. At the individual level, traffic safety is related to the absence of danger and experience of security*". Definition of the concept of "traffic system" is: "*Systems theory view used to describe the processes of the traffic system as dynamic and complex interactions between and among elements at various levels. The three main elements are usually identified as: the roadway infrastructure, the road-user, and the vehicle*" [5].

Basic dimensions of traffic safety. The easiest way road safety problem can be written in the following formula:

the traffic safety problem = exposure x risk x consequences.

This record was the first time used in a simple descriptive model. There were introduced the concept of "three basic dimensions of traffic safety":

1. risk = accidents/exposure) or risk = (injures/exposure);
2. consequences = (injures/accidents) or consequences = (fatalities/injures);
3. exposure (exposure of risk), defined by the measurement of exposure to accidents.

These three dimensions can be described on the Cartesian axles and unbind a cuboid, which will symbolized a quantitative picture of the road traffic safety [6, 7] (Fig. 2).

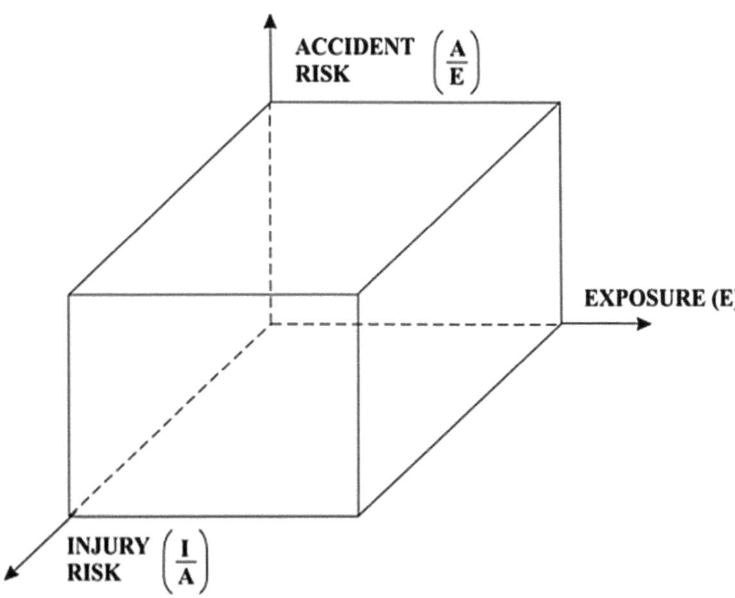

Fig. 2. Idea "three basic dimensions" of road traffic safety [6, 7]

3 Assessment of Road Safety in the EU

Road mobility comes at a high price in terms of lives lost: in 2014, slightly over 25 thousand persons lost their lives in road accidents within the EU-28 [8].

The total number of fatalities in road traffic accidents decreased by 45% between 2004 and 2014 (Fig. 1) at the level of the EU-28. A general trend towards fewer road traffic fatalities has long been observed in all countries in Europe. However, at the level of the EU, this downward trend has come to a standstill as the total number of fatalities registered in 2014 remained at the same level as in 2013. At regional level, there are significant variations in road safety [8].

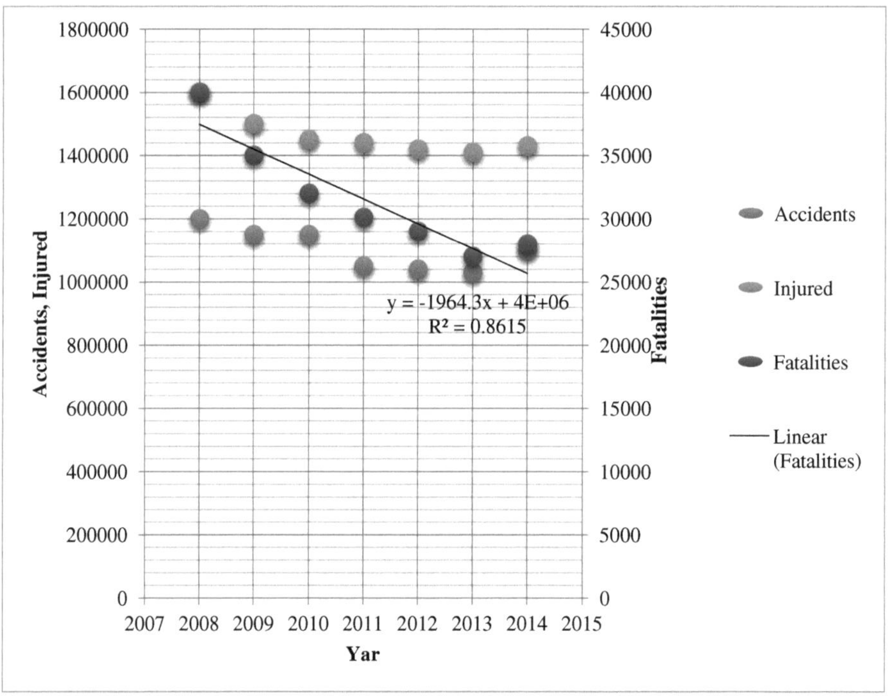

Fig. 3. Annual number of fatalities, injury accidents and injured people in the EU, 2005–2014

The steady decrease in the number of fatalities on Europe's roads appears to have come to a standstill. Nevertheless, despite the globally positive trend over the last decade (Fig. 3), the road death toll in 2014 is still more than 23 times the total number of fatalities in rail and air transport combined.

The number of road traffic fatalities in the various countries and regions depends on both structural differences:

- size of the country/region; composition,
- density and quality of the road network,

- characteristics of the population,
- socio-economic differences,
- characteristics of the vehicle stock,
- transit and tourist traffic,
- behavioural aspects etc. [8].

In order to determine "the real risk" in traffic amongst the different European countries, it is important to relate the number of accidents, injuries or fatalities to some indicator of traffic volume, i.e. exposure. By using actual accident figures alone it is almost impossible to compare the numbers for one country with the numbers for another country (Fig. 4).

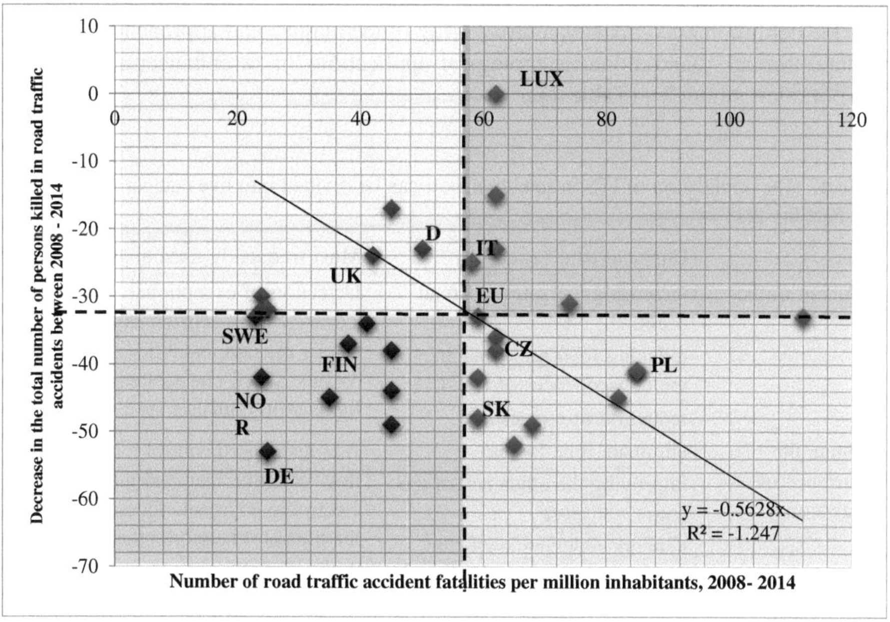

Fig. 4. The number of deaths and serious injuries to one million kilometers compared to annual average percentage changes from 2008–2014

4 Concept and Benefits of Its as a Tool of Management and Regulation of Road Transport

Information technology (IT) has transformed many industries, from education to health care to government, and is now in the early stages of transforming transportation systems. While many think improving a country's transportation system solely means building new roads or repairing aging infrastructures, the future of transportation lies not only in concrete and steel, but also increasingly in using IT.

IT enables elements within the transportation system—vehicles, roads, traffic lights, message signs, etc.—to become intelligent by embedding them with microchips and sensors and empowering them to communicate with each other through wireless technologies. In the leading nations in the world, ITS bring significant improvement in transportation system performance, including reduced congestion and increased safety and traveler convenience [9].

While the use of advanced technologies in transportation has been ongoing for many years, the creation of the ITS program has accelerated the pace of innovation and integration of technologies into the transportation system.

Intelligent Transportation Systems have been defined as:

"The application of advanced sensor, computer, electronics, and communication technologies and management strategies - in an integrated manner – to increase the safety and efficiency of the surface transportation system" **[10].**

This definition encompasses a broad array of techniques and approaches. This may be done through stand-alone technological applications or as enhancements to other transportation strategies (Table 1).

Table 1. The application of ITS in road transport can be divided into five categories in terms of their benefits [11]

Asset	Characteristic benefit	Most frequently used applications
Reducing traffic accidents	increasing the safety of road users	Detection of exceptional events and alarms systems ensure rapid response rescue operations CCTV systems to enforce speed changes and traffic lights Anti-collision systems
	reducing road accidents and the consequences of accidents	
	reduction of congestion	
Mitigate the occurrence of congestion	eliminate time losses	Adaptive speed control for the smooth flow of traffic Demand management: electronic payment, access control to the transport network efficiency of the transport network
	improving the efficiency of the entire transport	
	network reducing stress for drivers	
Performance operation	reduced operating costs	Automatic Automatic Vehicle location tracking freight operations management fleet Electronic fee
	increase productivity	
	improve the performance of transport process	
Environmental Protection	air pollution monitoring and prediction	Strategic demand management access control to highly concentrated urban communication networks for monitoring air pollution air quality information
Factors driving	comfort, safety trust in the transport network and transport system	Traffic information in real-time dynamic vehicle guidance to the target drive Vehicle Tracking security or emergency services

5 Design of Information System for Train Drivers

The guidance and information systems shall be regarded as affecting driver behavior data transmitted principally in two ways: for all vehicles of the traffic flow using information boards installed alongside roads or individually on display in the vehicle. The aim of the information system is to prevent dangerous situations in traffic and enable the driver to conveniently and timely decision on the basis of current information about traffic conditions on the roads. Published information on variable boards should focus on traffic closures, congestion, black ice and recommendations for detours. The operation of the traffic information of a considerable contributor to increase the safety and flow of traffic on the heavily indebted transport networks in major cities and on access roads. In the processing system, which has effectively operate in the traffic flow, the use of variable information boards (the PIT). These are an integral part of higher forms of management, information are not distributed manually from the control center operators on the basis of the health services in the relevant sections of the transport system of video surveillance or automatically based on other subsystems around management, such as exchange or traffic control systems tunnel. Other advantages PIT is the ability to re-report [12].

To achieve the effective use of electronic variable information signs within the road transport is possible in the case when road users are constantly provided with information about what is happening on the road. Generally, these instruments IDS can operate on the basis of four modes. Each mode has its advantages and specifications and it dependence on the traffic conditions on the relevant roads. Overview of modes and their functions are described in Table 2.

Table 2. Summary of the proposed arrangements and their function for variable electronic information board

Regime	Function
I.	**Basic Mode**, use in case of a traffic situation, without limitation, information mainly meteorological nature
II.	**Accident mode**, use the situation in case of a traffic limited to the guidance of the driver avoid the accident, preventing congestion
III.	**Work mode**, use for executing repair work on the road, narrowing to one lane,
IV.	**Additional mode**, use when treatment with strong gusty winds and torrential downpour at

Draft mode for variable electronic records information boards ensures that they achieved their full usability and as driver as road user will have constant access to information. A range of technological features (weather station, camera system, congestion detection, etc.) will benefit to ensure necessary information sharing.

One of the important information about traffic conditions is the surface temperature of the road. This is because as drivers as road users respond better to risk with the rapid changes in driving conditions such as snow, ice and rain. Mainly this is the reason that these conditions greatly affect the braking distance of the vehicle at a given speed.

The braking distance is considered as one of the decisive factors playing the important roles in the incident in road transport.

Braking distance (on a different road surface), including other factors makes to represent an approximate extension of 30%. It is therefore important that road users adjust their speed to the nature of the road, not to endanger themselves and other road users. Even with this reason, the proposed electronic guide sign consider the surface of the road and as a consequence it will recommend the traveling speed for the communication.

Early studies on the relationship between speed and accident showed that vehicles travelling more slowly than the average speed for the road, as well as those travelling faster, had an above–average risk of accident involvement, but that the severity of accident increased more than proportionally to speed. These early findings are consistent with subsequent findings that, on roads of a given type, injury accident rate, severe injury (including fatal) accident rate and fatal accident rate increase roughly as the powers of the mean traffic speed as illustrated in Figs. 5 and 6 [13].

Fig. 5. Example variable electronic bulletin board - basic mode - display of the max. velocity and the meteorological information.

Nilsson concluded that the number of accidents with serious injuries might be more affected by the change in the speed than the total number of accidents [13] (Fig. 7).

Accident rate and change rate are positively correlated. However, there is no logical connection between the change in speed and accidents. In addition, the changes do not affect the speed of the consequences arising from the occurrence of the accident. Second, the rate of increase in the absolute logical accidents. It shortens the time the driver to be available in critical situations for observation, decision-making and avoid. The increase in the absolute velocity in contrast to the rate of change increases, almost always increases the forces on impact. Consequently, the increase in speed increases the damage to the vehicles and the severity of injury to the crew.

Fig. 6. Example variable electronic bulletin board - basic mode - information about the next rest area.

In the event of an emergency on certain roads comes the option of driving through variable Electronic Bulletin Board (Accident mode). In general, when the information system is faced with three options for the management of existing vehicles:

1. Where using light signals, signals the vehicles are stopped.
2. At variable traffic signs, which subsequently will limit the speed of vehicles.
3. Option is the effect on traffic flows in the sense that the vehicles are moved to another (Diversionary) route.

While it is recognized that the driver must get to your destination, even different route, which may appear from the perspective of individual circumstance. Source and destination remains unchanged. It is mostly text messages like e.g.: "THE INTERNATIONAL AIRPORT 60 min ACCIDENT", and so on. It therefore only the driver's willingness to accept the information, and also for smoothing out the alternative route, since it is characterized in the special road markings, but can be provided as information for example "DIRECTION HRADEC - ONLY TOUR EDEN" [12].

Variable electronic records information boards can significantly affect the quality of use of transport networks. In Fig. 1 is a communication network made up of nodes K0 and K3 and the lines connecting the nodes. In the standard mode of transport using vehicles marked with the main communication route 1. This route is shorter and more straightforward than Route 2 at any given time is the main communication overloaded 2000 vehicles • h^{-1}, or a maximum capacity C1max = 1700 vehicles • h^{-1} yet the alternative route with 860 vehicles • h^{-1} well below its maximum capacity C2 = 1300 vehicles • h^{-1}. The real situation is then such that the main route slowly moving successive columns of vehicles in an unstable mode while driving alternative route is completely smooth.

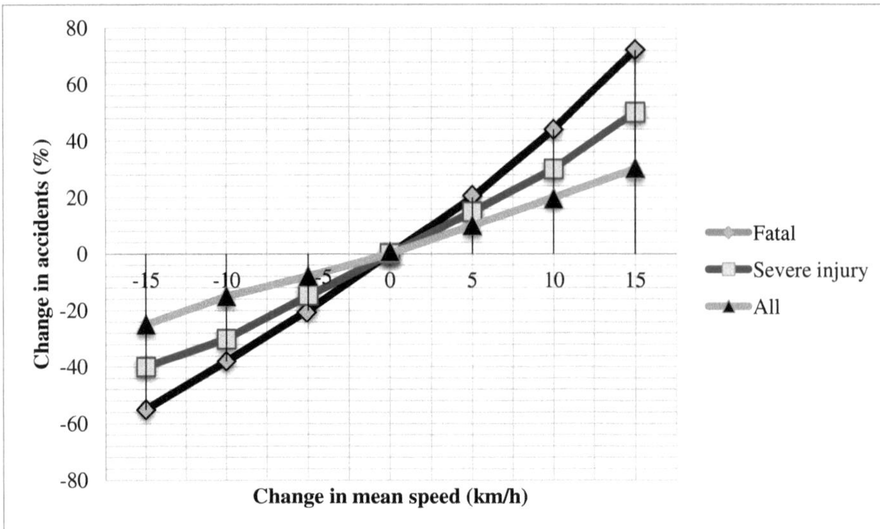

Fig. 7. Effects of mean speed on the number of accidents according to the Swedish model in case where the initial mean speed is 80 km/h.

As the prediction of the traffic on arrival and show that the intensities at the entrance for at least 15 min. change is activated to warn the driver information system to congestion on routes 1 and recommending use a route second [12] (Fig. 8).

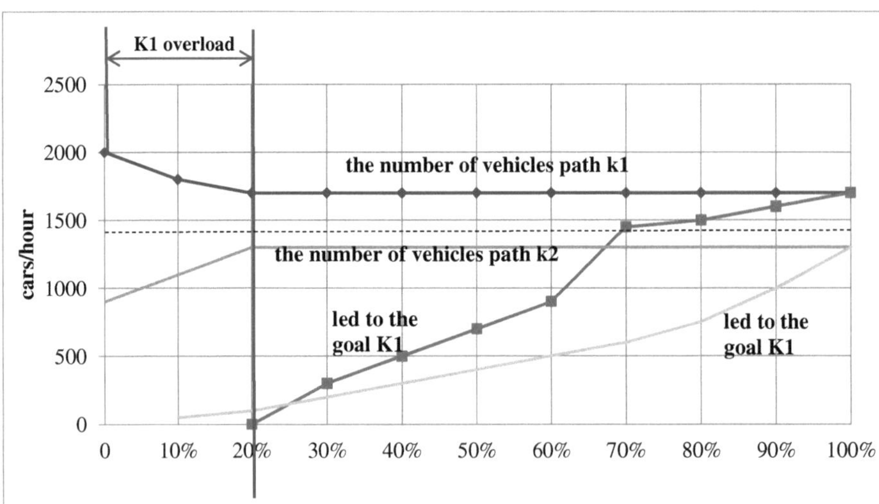

Fig. 8. Schematic representation of a communication network

The aim of the management process is to divide vehicles in relation to the capacity of communication. In that case, it is sufficient that 14% of drivers accepted alternative route and thereby achieved a balance in uniform loads, both communications.

Prerequisite, however, continuous measurement, prediction and evaluated parameters of traffic flow, because there must be the opposite extreme, which would mean instability traffic the alternative route [12].

Within the next two proposed scheme and additional work. Work mode will operate as long as a given road section will be carried out repairs, maintenance case. In most cases the shutdown of one lane and the speed is reduced. Through all the e variable information boards get the information to road users in time, allowing for the inclusion of vehicles to the transit through the lane while its attention to carrying out repairs. For feed mode, it will operate as long as the unlikely traffic obstruction on roads where restrictions would cause floods and floods, hail, storms and so on.

6 Summary

Intelligent transport systems are very effective and powerful tool. As for the traffic as for the state of emergency. Because of continuing evolution of vehicles, is necessary that the road infrastructure is adapted for it. The paper described the safety of road transport and the assessment of the road transport. The paper is focused on the possibility of using intelligent transport system. An important idea of intelligent transport systems is that these systems can not be bought, but must be efficient buildings. Efficient building of intelligent transport system brings the safety increase of the critical infrastructure elements in the sector of transport.

References

1. Rumar, K., Fleury, D., Lind, V., Berry, J.: Intelligent transportation systems and road safety. European Transport Safety Council Extracts from this Publication may be Reproduced with the Permission of ETSC, Brusel (1999). ISBN: 90-76024-05-7
2. Rumar, K.: Research for improved road safety. Paper at AA Foundation 10th Anniversary Conference, Birmingham, UK, 11 September 1996
3. Al-Haji, G.: Road safety development index (RSDI). theory, philosophy and practice. Linköping Studies in Science and Technology. Dissertation No. 1100. Department of Science and Technology, Linköping University, SE-601 74, Norrköping, Sweden 2007
4. Lederer, J.: Safety science in aviation. In: Proceedings of 1st World Congress on Safety Science, Köln (1990)
5. Arche, J.: Methods for the assessment and prediction of traffic safety at Urban intersections and their application in micro-simulation modelling (Academic Thesis). Royal Institute of Technology, Stockholm (2004). http://www.ctr.kth.se/publications/ctr2004_15.pdf
6. Rumar, K.: Transport safety visions, targets and strategies: beyond 2000. European Transport Safety Council, Brussels (1999). http://www.etsc.eu/oldsite/etsl1.pdf
7. 2RSP: Road safety principles and models: review of descriptive, predictive, risk and accident consequence models. OCDE/GD (97)153, Paris (1997). Public in Lukasik, Z., Szymanek, A., – Safety and risk in read traffic: selected problems
8. Eurostat – Statistics explained. Road safety statistics at regional level (2016). http://ec.europa.eu/eurostat/statistics-explained/index.php/Road_safety_statistics_at_regional_level

9. Ezell, S.: Intelligent Transportation Systems. Explaining International IT Application Leadership. The Information Technology a Innovation Foundation (2010). http://www.itif.org/files/2010-1-27-ITS_Leadership.pdf
10. Federal Highway Administration: Integrating intelligent transportation systems within the transportation planning process (1998). https://www.fhwa.dot.gov/publications/research/operations/its/98048/interimhb.pdf
11. Majerčák, P.: Prínosy a koncepcia Inteligentných dopravných systémov ako nástroja riadenia a regulovania dopravy v rámci Slovenskej republiky [on line]. Elektronický portál - Logistický monitor - Internetové noviny pre rozvoj logistiky na Slovensku (2014). ISSN: 1336-5851
12. Přibyl, P., Svítek, M.: Inteligentní Dopravní Systémy, BEN, Praha (2001). 248-255 s. ISBN 80-7300-029-6
13. VTT communities a Infrastructure. MASTER managing speed of traffic on European roads. Final Report for Publication (1998). http://www.transport-research.info/sites/default/files/project/documents/master.pdf

Author Index

A. Kravcov et al. (eds.), *Durability of Critical Infrastructure, Monitoring and Testing*, Lecture Notes in Mechanical Engineering, DOI 10.1007/978-981-10-3247-9

MIX
Papier aus verantwortungsvollen Quellen
Paper from responsible sources
FSC® C105338

If you have any concerns about our products,
you can contact us on
ProductSafety@springernature.com

In case Publisher is established outside the EU,
the EU authorized representative is:
Springer Nature Customer Service Center GmbH
Europaplatz 3, 69115 Heidelberg, Germany

Printed by Libri Plureos GmbH
in Hamburg, Germany